D0881915

CRISC™
REVIEW MANUAL
2015

Certified in Risk
and Information
Systems Control™

An ISACA® Certification

ISACA®

With more than 115,000 constituents in 180 countries, ISACA *(www.isaca.org)* helps business and IT leaders build trust in, and value from, information and information systems. Established in 1969, ISACA is the trusted source of knowledge, standards, networking, and career development for information systems audit, assurance, security, risk, privacy and governance professionals. ISACA offers the Cybersecurity Nexus™, a comprehensive set of resources for cybersecurity professionals, and COBIT®, a business framework that helps enterprises govern and manage their information and technology. ISACA also advances and validates business-critical skills and knowledge through the globally respected Certified Information Systems Auditor® (CISA®), Certified Information Security Manager® (CISM®), Certified in the Governance of Enterprise IT® (CGEIT®) and Certified in Risk and Information Systems Control™ (CRISC™) credentials. The association has more than 200 chapters worldwide.

Disclaimer

ISACA has designed and created *CRISC™ Review Manual 2015* primarily as an educational resource to assist individuals preparing to take the CRISC certification exam. It was produced independently from the CRISC exam and the CRISC Certification Committee, which has had no responsibility for its content. Copies of past exams are not released to the public and were not made available to ISACA for preparation of this publication. ISACA makes no representations or warranties whatsoever with regard to these or other ISACA publications assuring candidates' passage of the CRISC exam.

Reservation of Rights

ISACA

3701 Algonquin Road, Suite 1010
Rolling Meadows, IL 60008 USA
Phone: +1.847.253.1445
Fax: +1.847.253.1443
Email: *info@isaca.org*
Web site: *www.isaca.org*

Participate in the ISACA Knowledge Center: *www.isaca.org/knowledge-center*
Follow ISACA on Twitter: *https://twitter.com/ISACANews*
Join ISACA on LinkedIn: ISACA (Official), *http://linkd.in/ISACAOfficial*
Like ISACA on Facebook: *www.facebook.com/ISACAHQ*

ISBN 978-1-60420-590-9
CRISC™ Review Manual 2015
Printed in the United States of America

CRISC REVIEW MANUAL 2015

ISACA is pleased to offer the 2015 (5th) edition of the *CRISC™ Review Manual*. The purpose of the manual is to provide CRISC candidates with information and references to assist in the preparation and study for the Certified in Risk and Information Systems Control™ (CRISC) exam.

The content in the manual has been substantially updated and reformatted to improve the study process. Most of the changes made were to recognize and map to the new domains and task and knowledge statements that resulted from the new CRISC job practice analysis. Further details regarding the new job practice can be found in the section on page v titled NEW–CRISC Job Practice and can be viewed at *www.isaca.org/criscjobpractice* and in the *ISACA Exam Candidate Information Guide* at *www.isaca.org/examguide*. **The exam is based on the task and knowledge statements in the job practice.** The development of the task and knowledge statements involved thousands of CRISCs and other industry professionals worldwide who served as committee members, focus group participants, subject matter experts and survey respondents.

The *CRISC™ Review Manual* is updated annually to keep pace with rapid changes in the identification, assessment, response, monitoring and reporting of risk and information systems (IS) controls. As with previous manuals, the 2015 edition is the result of contributions from many qualified authorities who have generously volunteered their time and expertise. We respect and appreciate their contributions and hope their efforts provide extensive educational value to CRISC manual readers.

Your comments and suggestions regarding this manual are welcome. After taking the exam, please take a moment to complete the online questionnaire *(www.isaca.org/studyaidsevalutation)*. Your observations will be invaluable for the preparations of the 2016 edition of the *CRISC™ Review Manual*.

The sample questions contained in this manual are designed to depict the type of questions typically found on the CRISC exam and to provide further clarity to the content presented in this manual. The CRISC exam is a practice-based exam. Simply reading the reference material in this manual will not properly prepare candidates for the exam. The self-assessment questions are included for guidance only. Scoring results do not indicate future individual exam success.

Certification has resulted in a positive impact on many careers, including worldwide recognition for professional experience and enhanced knowledge and skills. The Certified in Risk and Information Systems Control™ certification (CRISC™, pronounced "see-risk") is designed for IT and business professionals who have hands-on experience with risk identification, risk assessment, risk response, and risk and IS control monitoring and reporting. ISACA wishes you success with the CRISC exam.

ACKNOWLEDGMENTS

The *CRISC™ Review Manual 2015* is the result of collective efforts of many volunteers. ISACA members from throughout the globe participated, generously offering their talent and expertise. This international team exhibited a spirit and selflessness that has become the hallmark of contributors to ISACA manuals. Their participation and insight are truly appreciated.

Special thanks go to Kevin M. Henry, CISA, CISM, CRISC, CISSP, Mile2, Canada, who worked on the 2015 edition of the *CRISC™ Review Manual*.

Expert Reviewers
Ronald D. Burns, CRISC, Bank of Internet Federal Bank, USA
Alvaro Rodrigo Cayul, CISM, CRISC, Chile
Bobbette R. Fagel, CISA, CISM, CRISC, CISSP, MTA, Capital Compliance Services, USA
Shawna M. Flanders, CISA, CISM, CRISC, CSSGB, SSBB, Business - Technology Guidance Associates, LLC, USA
Robert Thomas Hanson, CISA, CISM, CRISC, Australia
W. Noel Haskins-Hafer, CISA, CISM, CGEIT, CRISC, CIA, CFE, CRMA, SCPM, Intuit, Inc., USA
Ken Hendrie, CISA, CISM, CGEIT, CRISC, GCIH, ITIL, PRINCE2, BAE Systems Applied Intelligence, Australia
Ramaswami Karunanithi, CISA, CGEIT, CRISC, CA, CAMS, CBCI, CCSA, CFE, CFSA, CGMA, CIA, CMA,
    CPA, CRMA, FCS, PMP, New South Wales Government, Australia
Ravikumar Ramachandran, CISA, CISM, CGEIT, CRISC, CAP, CEH, CFE, CHFI, CIA, CISSP-ISSAP, CRMA,
    ECSA, FCMA, PMP, SSCP, Hewlett-Packard India Sales Pvt. Ltd, India
Jan van Prooijen, CISA, CRISC, CISSP, HP ESS, The Netherlands

ISACA has begun planning the 2016 edition of the *CRISC™ Review Manual*. Volunteer participation drives the success of the manual. If you are interested in becoming a member of the select group of professionals involved in this global project, we want to hear from you. Please email us at *studymaterials@isaca.org*.

NEW—CRISC JOB PRACTICE

BEGINNING IN 2015, THE CRISC EXAM WILL TEST THE NEW CRISC JOB PRACTICE.

An international job practice analysis is conducted at least every five years or sooner to maintain the validity of the CRISC certification program. A new job practice forms the basis of the CRISC exam beginning in June 2015.

The primary focus of the job practice is the current tasks performed and the knowledge used by CRISCs. By gathering evidence of the current work practice of CRISCs, ISACA is able to ensure that the CRISC program continues to meet the high standards for the certification of professionals throughout the world.

The findings of the CRISC job practice analysis are carefully considered and directly influence the development of new test specifications to ensure that the CRISC exam reflects the most current best practices.

The new 2015 job practice reflects the areas of study to be tested and is compared below to the previous job practice.

Previous CRISC Job Practice	New 2015 CRISC Job Practice
Domain 1: Risk Identification, Assessment and Evaluation (31%) Domain 2: Risk Response (17%) Domain 3: Risk Monitoring (17%) Domain 4: Information Systems Control Design and Implementation (17%) Domain 5: Information Systems Control Monitoring and Maintenance (18%)	**Domain 1: IT Risk Identification (27%)** **Domain 2: IT Risk Assessment (28%)** **Domain 3: Risk Response and Mitigation (23%)** **Domain 4: Risk and Control Monitoring and Reporting (22%)**

Page intentionally left blank

About This Manual

OVERVIEW

The *CRISC™ Review Manual 2015* is a reference guide designed to assist candidates in preparing for the CRISC examination. **The manual is one source of preparation for the exam, but should not be thought of as the only source nor viewed as a comprehensive collection of all the information and experience that are required to pass the exam.** No single publication offers such coverage and detail.

As candidates read through the manual and encounter topics that are new to them or ones in which they feel their knowledge and experience are limited, additional references should be sought. The examination will be composed of questions testing the candidate's technical and practical knowledge, and ability to apply the knowledge (based on experience) in given situations.

ORGANIZATION OF THIS MANUAL

The *CRISC™ Review Manual 2015* is divided into four chapters covering the CRISC domains tested on the exam in the percentages listed below:

Domain 1	IT Risk Identification	27 percent
Domain 2	IT Risk Assessment	28 percent
Domain 3	Risk Response and Mitigation	23 percent
Domain 4	Risk and Control Monitoring and Reporting	22 percent

> **NOTE:** Each chapter defines the tasks that CRISC candidates are expected to know how to do and includes a series of knowledge statements required to perform those tasks. These constitute the current practices for the IS risk professional. **The detailed CRISC job practice can be viewed at *www.isaca.org/criscjobpractice*. The exam is based on these task and knowledge statements.**

The manual has been developed and organized to assist in the study of these areas. Exam candidates should evaluate their strengths, based on knowledge and experience, in each of these areas.

FORMAT OF THIS MANUAL

Each of the four chapters of the *CRISC™ Review Manual 2015* is divided into two sections for focused study. Section One is an overview that provides:
• A definition for each domain
• Learning objectives for each domain
• A listing of the task and knowledge statements for each domain
• Sample self-assessment questions, answers and explanations
• Suggested resources for further study

Section Two consists of reference material and content that supports the knowledge necessary to perform the task statements. Material included is pertinent to CRISC candidates' knowledge and/or understanding when preparing for the CRISC certification exam.

The structure of the content includes numbering to identify the chapter where a topic is located and headings of the subsequent levels of topics addressed in the chapter (i.e., 1.3.1 Risk Culture is a subtopic of 1.3 Risk Culture and Communication in chapter 1). Relevant content in a subtopic is bolded for specific attention.

Understanding the material is a barometer of the candidate's knowledge, strengths and weaknesses, and is an indication of areas in which the candidate needs to seek reference sources over and above this manual. However, written material is not a substitute for experience. **CRISC exam questions will test the candidate's practical application of this knowledge.** The self-assessment questions in the first section of each chapter assist in understanding how a CRISC question could be presented on the CRISC exam and should not be used independently as a source of knowledge. Self-assessment questions should not be considered a measurement of the candidate's ability to answer questions correctly on the CRISC exam for that area. The questions are intended to familiarize the candidate with question structure, and may or may not be similar to questions that will appear on the actual examination. The reference material includes other publications that could be used to further acquire and better understand detailed information on the topics addressed in the manual.

A glossary is included at the end of the manual and contains terms that apply to the material included in the chapters. Also included are terms that apply to related areas not specifically discussed. The glossary is an extension of the text in the manual and can, therefore, be another indication of areas in which the candidate may need to seek additional references.

Although every effort is made to address the majority of information that candidates are expected to know, not all examination questions are necessarily covered in the manual, and candidates will need to rely on professional experience to provide the best answer.

Throughout the manual, "association" refers to ISACA, formerly known as Information Systems Audit and Control Association, and "institute" or "ITGI®" refers to the IT Governance Institute®. Also, please note that the manual has been written using standard American English.

NOTE: The *CRISC™ Review Manual 2015* is a living document. As the field of IT-related business risk management in information systems controls evolve, the manual will be updated to reflect such changes. Further updates to this document before the date of the exam may be viewed at *www.isaca.org/studyaidupdates.*

EVALUATION OF THIS MANUAL

ISACA continuously monitors the swift and profound professional, technological and environmental advances affecting the IS risk profession. Recognizing these rapid advances, the *CRISC™ Review Manual* is updated annually.

To assist ISACA in keeping abreast of these advances, please take a moment to evaluate the *CRISC™ Review Manual 2015.* Such feedback is valuable to fully serve the profession and future CRISC exam registrants.

To complete the evaluation on the web site, please go to *www.isaca.org/studyaidsevaluation.*

Thank you for your support and assistance.

ABOUT THE CRISC REVIEW QUESTIONS, ANSWERS & EXPLANATIONS MANUAL

Candidates may also wish to enhance their study and preparation for the exam by using the *CRISC™ Review Questions, Answers & Explanations Manual 2015* and the *CRISC™ Review Questions, Answers & Explanations Manual 2015 Supplement.*

The *CRISC™ Review Questions, Answers & Explanations Manual 2015* consists of 400 multiple-choice study questions, answers and explanations arranged in the domains of the current CRISC job practice. The questions in this manual appeared in the *CRISC™ Review Questions, Answers & Explanations Manual 2013* and in the 2013 and 2014 editions of the *CRISC™ Review Questions, Answers & Explanations Manual Supplement.*

The 2015 edition of the *Supplement* is the result of ISACA's dedication each year to create new sample questions, answers and explanations for candidates to use in preparation for the CRISC exam. Each year, ISACA develops 100 new review questions, using a strict process of review similar to that performed for the selection of questions for the CRISC exam by the CRISC Certification Committee. In the 2015 edition of the *Supplement*, the questions have been arranged in the proportions of the most recent CRISC job practice.

Questions in these publications are representative of the types of questions that could appear on the exam and include explanations of the correct and incorrect answers. Questions are sorted by the CRISC domains and as a sample test. These publications are ideal for use in conjunction with the *CRISC™ Review Manual 2015*. These manuals can be used as study sources throughout the study process or as part of a final review to determine where candidates may need additional study. It should be noted that these questions and suggested answers are provided as examples; they are not actual questions from the examination and may differ in content from those that actually appear on the exam.

ABOUT THE CRISC REVIEW QUESTIONS, ANSWERS & EXPLANATIONS DATABASE

Another study aid that is available is the CRISC™ Review Questions, Answers & Explanations Database – 12 Month Subscription. The online database consists of the 500 questions, answers and explanations included in the *CRISC™ Review Questions, Answers & Explanations Manual 2015* and the 2015 edition of the *Supplement*. With this product, CRISC candidates can quickly identify their strengths and weaknesses by taking random sample exams of varying length and breaking the results down by domain. Sample exams also can be chosen by domain, allowing for concentrated study, one domain at a time, and other sorting features such as the omission of previous correctly answered questions are available.

NOTE: When using the CRISC review materials to prepare for the exam, it should be noted that they cover a broad spectrum of IT-related business risk and IS control issues. **Again, candidates should not assume that reading these manuals and answering review questions will fully prepare them for the examination.** Since actual exam questions often relate to practical experiences, candidates should refer to their own experiences and other reference sources, and draw on the experiences of colleagues and others who have earned the CRISC designation.

Page intentionally left blank

Introduction to IT Risk Management

Introduction to IT Risk Management

Risk is defined as the combination of the probability of an event and its consequence. Often, risk is seen as an adverse event that can threaten an organization's assets or exploit vulnerabilities and cause harm. Several factors are considered when evaluating risk, such as the mission of the organization, assets, threat, vulnerability, likelihood and impact. These terms will be further explored in each of the chapters in this review manual.

GOVERNANCE AND RISK MANAGEMENT

Governance is the accountability for protection of the assets of an organization. The board of directors of the organization is accountable for governance, and the board entrusts the senior management team with the responsibility to manage the day-to-day operations of the organization in alignment with the strategic mandates approved by the board. Governance is applicable to all departments of the organization—financial accountability and oversight, operational effectiveness, legal and human resources compliance, social responsibility and governance of IT investment, operations, and control. Risk management is an important part of governance. Management requires accurate information to be able to correctly understand risk and address the circumstances that would indicate the need for risk mitigation.

Over the past decade, the term "governance" has moved to the forefront of business thinking in response to examples demonstrating the importance of good governance and, on the other end of the scale, global business mishaps. Corporate governance is the system by which organizations are evaluated, directed and controlled. The corporate governance of IT is the system by which the current and future use of IT is evaluated, directed and controlled. The objective of any governance system is to enable organizations to create value for their stakeholders. Consequently, value creation is a governance objective for any organization. Value creation is comprised of benefits realization, risk optimization and resource optimization. Risk optimization is, therefore, an essential part of any governance system and cannot be seen in isolation from benefits realization or resource optimization.

Governance answers four questions:
1. Are we doing the right things?
2. Are we doing them the right way?
3. Are we getting them done well?
4. Are we getting the benefits?

There is a clear distinction between governance and management. Management focuses on planning, building, running and monitoring within the directions set by the governance system to create value by achieving objectives. Risk management foresees the challenges to achieving these objectives and attempts to lower the chances and impacts of them occurring.

Exhibit 0.1 provides an overview of the risk governance structure.

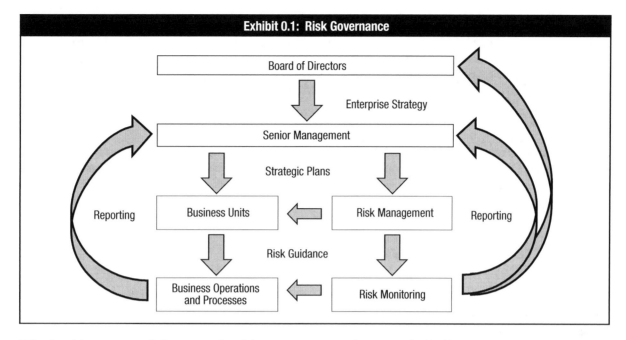

Exhibit 0.1: Risk Governance

Effective risk governance helps ensure that risk management practices are embedded in the enterprise, enabling it to secure optimal risk-adjusted return. Risk governance has four main objectives (as seen in **exhibit 0.2**):
1. Establish and maintain a common risk view.
2. Integrate risk management into the enterprise.
3. Make risk-aware business decisions.
4. Ensure that risk management controls are implemented and operating correctly.

Exhibit 0.2: Risk Governance Objectives	
Objective	**Description**
1. Establish and maintain a common risk view.	Effective risk governance establishes the common view of risk for the enterprise. This determines which controls are necessary to mitigate risk and how risk-based controls are integrated into business processes and information security. The risk governance function sets the tone of the business regarding how to determine an acceptable level of risk tolerance. Risk governance is a continuous life cycle that requires regular reporting and ongoing review. The risk governance function must oversee the operations of the risk management team.
2. Integrate risk management into the enterprise.	Integrating risk management into the enterprise enforces a holistic enterprise risk management (ERM) approach across the entire enterprise. It requires the integration of risk management into every department, function, system and geographic location. Understanding that risk in one department or system may pose an unacceptable risk to another department or system requires that all business processes be compliant with a baseline level of risk management. The objective of ERM is to establish the authority to require all business processes to undergo a risk analysis on a periodic basis or when there is a significant change to the internal or external environment.
3. Make risk-aware business decisions.	To make risk-aware business decisions, the risk governance function must consider the full range of opportunities and consequences of each such decision and its impact on the enterprise, society and the environment.
4. Ensure that risk management controls are implemented and operating correctly.	Governance requires oversight and due diligence to ensure that the enterprise is following up on the implementation and monitoring of controls to ensure that the controls are effective to mitigate risk and protect organizational assets.

THE CONTEXT OF IT RISK MANAGEMENT

Risk management is defined as the coordinated activities to direct and control an enterprise with regard to risk. In simple terms, risk can be considered as a challenge to achieving objectives. Therefore, risk management can be considered as the activity undertaken to foresee challenges and lower the chances of those challenges occurring and their impact. Effective risk management can also assist in maximizing opportunities.

International Organization for Standardization/International Electrotechnical Commission (ISO/IEC) 31000 states, "Risk is the effect of uncertainty on objectives. An effect is a deviation from the expected—positive and/or negative." However, ISO/IEC 27005 regards risk solely from a negative angle, stating "information security risk is the potential that a given threat will exploit vulnerabilities of an asset or group of assets and thereby cause harm to the organization." There are levels of risk, and the greater the risk, the higher the probability of loss.

Risk management starts with understanding the organization, but the organization is mostly a servant of the environment, or context, in which it operates. Assessing an organization's context includes evaluating the intent and capability of threats; the relative value of, and trust required in, assets (or resources); and the respective relationship of vulnerabilities that threats could exploit to intercept, interrupt, modify or fabricate data in information assets. Other factors that must be considered are the dependency of the organization on a supply chain (especially one based in another geographic region of the world), financing, debt and partners. Other considerations include:
• Vulnerability to changes in economic or political conditions
• Changes to market trends and patterns
• Emergence of new competition
• Impact of new legislation
• Existence of potential natural disaster
• Constraints caused by legacy systems and antiquated technology
• Strained labor relations and inflexible management

The strategy of the organization will drive the individual lines of business that make up the organization, and each line of business will develop information systems that support its business function. **Exhibit 0.3** illustrates how IT risk relates to overall risk of the organization.

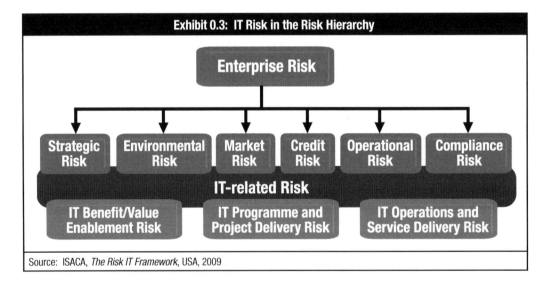

Exhibit 0.3: IT Risk in the Risk Hierarchy

Source: ISACA, *The Risk IT Framework*, USA, 2009

Risk is an influencing factor and must be evaluated at all levels of the organization—the strategic level, the business unit level and the information systems level. A properly managed risk framework addresses and takes into consideration the impact of risk at all levels and describes how a risk at one level may affect the other levels as well.

IT risk management is the implementation of a risk strategy that reflects the culture, appetite and tolerance levels of the senior management of the organization; considers technology and budgets; and addresses the requirements of regulation and compliance. In this way, an effective IT risk management strategy is critical to an organization's ability to execute its overall business strategy in an effective and efficient manner.

> There are several key parts of an IT risk management program. Different risk management methodologies use slightly different terms to describe the components of IT risk management. The CRISC candidate is not expected to be familiar with the details of each methodology, but should be familiar with the general concepts and process flows related to IT risk management.

IT risk management is a cyclical process, as shown in **exhibit 0.4**. The first step in the IT risk management process is the identification of IT risk, which includes determining the risk context and risk framework, and the process of identifying and documenting risk. The risk identification effort should result in the listing and documentation of risk. This step aligns with the next phase of the IT risk management process: IT risk assessment. The effort to assess risk, including the prioritization of risk, will provide management with the data required for consideration as a key factor in the next phase, risk response and mitigation. Risk response and mitigation addresses the risk appetite and tolerance of the organization and the need to find cost-effective ways to address risk. The final phase of IT risk management is risk and control monitoring and reporting. In this phase, controls and risk management efforts, as well as the current risk state, are monitored and results are reported back to senior management, who will determine the need to return to any of the previous phases of the process.

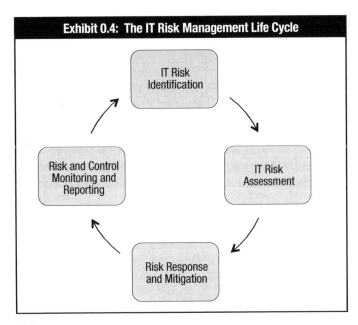

Exhibit 0.4: The IT Risk Management Life Cycle

The IT risk management process is based on the complete cycle of all the elements. A failure to perform any one of the phases in a complete and thorough manner will result in an ineffective risk management process. A failure in any step of the cycle may cause a deficiency that will affect the other phases. As with all life cycles, the process continues with refinement, adaptation and a focus on continuous improvement and maturity. The more often the risk management life cycle is repeated and continuously improved, the more effective the IT risk management effort will be, and consistent results will be obtained.

IMPORTANCE OF IT RISK MANAGEMENT

IT risk management, especially organizationwide risk management, is a valuable part of the governance and effective management of the organization. It addresses the uncertainty of what may, or may not, happen and the measures taken to address the effects of an event if it were to take place. The benefits of IT risk management include:

• Better oversight of organizational assets
• Minimized loss
• Identification of threats, vulnerabilities and risk
• Prioritization of risk response efforts
• Legal and regulatory compliance
• Increased likelihood of project success
• Improved performance and the ability to attain business goals
• Increased confidence of stakeholders (owners, customers, employees, regulators, etc.)
• Creation of a risk-aware culture
• Better incident and business continuity management
• Improved controls
• Better monitoring and reporting
• Improved decision making
• Ability to meet business objectives

BUSINESS RISK VERSUS IT RISK

Risk is a critical part of business. Unless a business is willing to take a risk, it will not be able to realize the benefits associated with risk. However, taking too much risk may lead to increased likelihood of failure of the business and loss of investment. Every business faces the decision of how much risk to take and what opportunities to forego. This is a decision that reflects the risk acceptance level of the senior management team.

The risk practitioner is primarily concerned with IT risk, which is a subset of business risk. This requires the risk practitioner to understand the risk culture of an organization and let that drive the IT risk strategy. When calculating IT risk, the risk practitioner must be careful not to calculate risk solely from the perspective of the impact of the risk on IT and to ensure that both the technical and nontechnical elements of risk have been considered. An IT system failure has an impact on the IT department, but it may have a much greater impact on the business supported by the IT system than on IT alone.

> **For the purposes of this manual, "risk practitioner" will refer to the "IT risk practitioner."**

The role of IT is to serve the business, and the need to support business requirements is IT's primary responsibility. The business does not exist so that the organization can have an IT department; the IT department exists to help the business meet its mission and goals. When this is not understood, conflicts arise between IT and the business units. In some cases, this leads to a decision to outsource IT altogether. The relationship between IT and the business is a critical success factor (CSF) that must be considered by the risk practitioner because it will affect the ability of the organization to achieve its objectives by being flexible in adopting new technologies and will result in lower resistance from the employees to comply with security controls. The IT risk strategy is, therefore, driven by, and reflective of, the business risk strategy. Any factor that threatens to undermine the relationship between the business and IT is a risk to value creation.

The risk practitioner may also examine areas such as business continuity, IS audit, information security, IS controls, projects and change.

Risk and Business Continuity

IT risk management is closely linked with business continuity, and IT risk assessment is often a precursor to a business impact analysis (BIA). Business continuity planning is concerned with the preservation of critical business functions and the ability of the organization to survive an adverse event that may impact the ability of the organization to meet its mission and goals.

In many ways, business continuity starts where risk management ends. Through IT risk management, the organization attempts to reduce all IT risk to an acceptable level. Although the controls and efforts of IT risk management may not prevent a failure, the risk practitioner works with the incident management and business continuity teams to identify possible threats and put in place the mechanism to detect, contain and recover from an adverse event if it should happen. The risk is that the business continuity plan (BCP) may not be adequate or accurate, thereby leading to a failure to recover effectively from an incident.

IT Risk and IS Audit

IS audit is an important part of corporate governance; audit provides assurance to management on the effectiveness of the IS control framework, IT risk management and compliance. In a world of increasing legislation, government oversight and media scrutiny, organizations must diligently demonstrate an adequate control environment and risk management. IS audit should be conducted by objective, skilled and independent personnel who are able to assess risk, identify vulnerabilities, document findings and provide recommendations on how to address audit issues.

The risk associated with IS audit is often related to the competence of the IS audit personnel. An audit is a methodical and structured review that requires the IS auditor to be competent and knowledgeable about the area being audited. If the IS auditor is not familiar with the technology being used, the significance of operating conditions or the requirements of the organization, the audit may be inaccurate and provide limited value.

Another IS audit risk is associated with independence of the audit. Senior management is often involved in the creation of the IS audit plan, and in a situation where management is involved in inappropriate activity, they may restrict the ability of IS audit to perform their duties effectively.

IT Risk and Information Security

Information security is usually based on risk. The National Institute of Standards and Technology (NIST) states that an organization must provide risk-based, cost-effective controls. The effectiveness of the information security program is based on a foundation of thorough and accurate IT risk management. If the IT risk management activity is not conducted properly, then the information security controls are almost certain to be incorrectly designed, poorly implemented and improperly operated.

IT risk drives the selection of controls and justifies the choice and operation of a control. Every control should be traceable back to a specific IT risk that the control is designed to mitigate. The risk practitioner should be able to demonstrate the purpose of each control and explain the reasoning behind the selection and enforcement of the control.

Control Risk

A control is chosen to mitigate a risk, but if the control is not operating correctly then the control may not prevent a failure or compromise. The selection of the wrong control, the incorrect configuration of the control, the improper operation of the control, the failure to monitor and review the control, or the inadequacy of the control to address new threats may all present the risk of control failure.

Project Risk

Many, if not most, projects fail. Numerous studies of IT projects have indicated that a majority of IT projects could be considered a failure.[1] There are several ways to determine failure: over budget, over the allotted time schedule, or failure to meet customer needs and expectations. The failure of an IT project may pose a significant risk to an organization including a loss of market share; failure to seize new opportunities; and impact on customers, shareholders and employee confidence. Identifying the risk associated with a project and managing that risk may result in a higher incidence of project success and stakeholder satisfaction.

Change Risk

After an information system or business process goes into operation, the risk changes and new risk emerges. Changes in technology, regulations, business processes, functionality, architecture and users may affect the risk associated with a system that is in operation. There also may be changes in the operational environment that result in the controls that were originally designed becoming ineffective.

An important task of the risk practitioner is to manage risk on a continuous basis and to be aware of emerging risk, new threats, new technologies, changes in culture and increased legislation, and/or regulation. All of these changes may affect the risk posture of the organization and result in a new level of risk not adequately addressed in earlier risk identification efforts.

SUMMARY

This section provided an overview of the areas of IT risk that will be addressed by the risk practitioner. There are many variables that a risk practitioner must consider and many decisions that a risk practitioner must make, but the success of the IT risk management effort is usually based on having an organizationwide perspective of risk, following a structured methodology and gathering the correct information. It is through the success of the IT risk management effort that a risk practitioner will be able to add value, recommend appropriate controls, and report status of the risk profile to management and all relevant stakeholders.

ENDNOTES

[1] McManus, John; Trevor Wood-Harper; "A Study in Project Failure," BCS, The Chartered Institute for IT, June 2008, *www.bcs.org/content/ConWebDoc/19584*

Chapter 1: IT Risk Identification

Section One: Overview

Section Two: Content

Section One: Overview

DOMAIN DEFINITION

Identify the universe of IT risk to contribute to the execution of the IT risk management strategy in support of business objectives and in alignment with the enterprise risk management (ERM) strategy.

LEARNING OBJECTIVES

The objective of this domain is to ensure that the CRISC candidate has the knowledge necessary to:
• Identify relevant standards, frameworks and practices
• Apply risk identification techniques
• Distinguish between threats and vulnerabilities
• Identify relevant stakeholders
• Discuss risk scenario development tools and techniques
• Explain the meaning of key risk management concepts, including risk appetite and risk tolerance
• Describe the key elements of a risk register
• Contribute to the creation of a risk awareness program

CRISC EXAM REFERENCE

This domain represents 27 percent of the CRISC exam (approximately 41 questions).

TASK AND KNOWLEDGE STATEMENTS

TASKS

There are seven tasks within this domain that a CRISC candidate must know how to perform. These relate to IT risk identification.

T1.1   Collect and review information, including existing documentation, regarding the organization's internal and external business and IT environments to identify potential impacts of IT risk to the organization's business objectives and operations.

T1.2   Identify potential threats and vulnerabilities to the organization's people, processes and technology to enable IT risk analysis.

T1.3   Develop a comprehensive set of IT risk scenarios based on available information to determine the potential impact to business objectives and operations.

T1.4   Identify key stakeholders for IT risk scenarios to help establish accountability.

T1.5   Establish an IT risk register to help ensure that identified IT risk scenarios are accounted for and incorporated into the enterprisewide risk profile.

T1.6   Identify risk appetite and tolerance defined by senior leadership and key stakeholders to ensure alignment with business objectives.

T1.7   Collaborate in the development of a risk awareness program, and conduct training to ensure that stakeholders understand risk and to promote a risk-aware culture.

KNOWLEDGE STATEMENTS

The CRISC candidate should be familiar with the task statements relevant to each domain in the CRISC job practice. The tasks are supported by 41 knowledge statements that delineate each of the areas in which the risk practitioner must have a good understanding in order to perform the tasks. Many knowledge statements support tasks that cross domains.

The CRISC candidate should have knowledge of:
1. Laws, regulations, standards and compliance requirements
2. Industry trends and emerging technologies
3. Enterprise systems architecture (e.g., platforms, networks, applications, databases and operating systems)
4. Business goals and objectives
5. Contractual requirements with customers and third-party service providers

6. Threats and vulnerabilities related to:
   6.1. Business processes and initiatives
   6.2. Third-party management
   6.3. Data management
   6.4. Hardware, software and appliances
   6.5. The system development life cycle (SDLC)
   6.6. Project and program management
   6.7. Business continuity and disaster recovery management (DRM)
   6.8. Management of IT operations
   6.9. Emerging technologies
7. Methods to identify risk
8. Risk scenario development tools and techniques
9. Risk identification and classification standards, and frameworks
10. Risk events/incident concepts (e.g., contributing conditions, lessons learned, loss result)
11. Elements of a risk register
12. Risk appetite and tolerance
13. Risk analysis methodologies (quantitative and qualitative)
14. Organizational structures
15. Organizational culture, ethics and behavior
16. Organizational assets (e.g., people, technology, data, trademarks, intellectual property) and business processes, including enterprise risk management (ERM)
17. Organizational policies and standards
18. Business process review tools and techniques
19. Analysis techniques (e.g., root cause, gap, cost-benefit, return on investment [ROI])
20. Capability assessment models and improvement techniques and strategies
21. Data analysis, validation and aggregation techniques (e.g., trend analysis, modeling)
22. Data collection and extraction tools and techniques
23. Principles of risk and control ownership
24. Characteristics of inherent and residual risk
25. Exception management practices
26. Risk assessment standards, frameworks and techniques
27. Risk response options (i.e., accept, mitigate, avoid, transfer) and criteria for selection
28. Information security concepts and principles, including confidentiality, integrity and availability of information
29. Systems control design and implementation, including testing methodologies and practices
30. The impact of emerging technologies on design and implementation of controls
31. Requirements, principles, and practices for educating and training on risk and control activities
32. Key risk indicators (KRIs)
33. Risk monitoring standards and frameworks
34. Risk monitoring tools and techniques
35. Risk reporting tools and techniques
36. IT risk management best practices
37. Key performance indicator (KPIs)
38. Control types, standards, and frameworks
39. Control monitoring and reporting tools and techniques
40. Control assessment types (e.g., self-assessments, audits, vulnerability assessments, penetration tests, third-party assurance)
41. Control activities, objectives, practices and metrics related to:
   41.1. Business processes
   41.2. Information security, including technology certification and accreditation practices
   41.3. Third-party management, including service delivery
   41.4. Data management
   41.5. The system development life cycle (SDLC)
   41.6. Project and program management
   41.7. Business continuity and disaster recovery management (DRM)
   41.8. IT operations management
   41.9 The information systems architecture (e.g., platforms, networks, applications, databases and operating systems)

SELF-ASSESSMENT QUESTIONS

1-1   Which of the following business requirements **BEST** relates to the need for resilient business and information systems processes?

   A. Effectiveness
   B. Confidentiality
   C. Integrity
   D. Availability

1-2   The preparation of a risk register begins in which of the following risk management processes?

   A. Risk response planning
   B. Risk monitoring and control
   C. Risk identification
   D. Risk management strategy planning

1-3   Shortly after performing the annual review and revision of corporate policies, a risk practitioner becomes aware that a new law may affect security requirements for the human resources system. The risk practitioner should:

   A. analyze in detail how the law may affect the enterprise.
   B. ensure that necessary adjustments are implemented during the next review cycle.
   C. initiate an *ad hoc* revision of the corporate policy.
   D. notify the system custodian to implement changes.

1-4   Which of the following choices is a function of risk governance?

   A. Establish business objectives.
   B. Develop the business strategy.
   C. Create the risk management strategy.
   D. Align risk management and business strategies.

1-5   Which of the following choices provides the **BEST** view of risk management?

   A. An interdisciplinary team
   B. A third-party risk assessment service provider
   C. The enterprise's IT department
   D. The enterprise's internal compliance department

1-6   Which of the following choices is a **PRIMARY** consideration when developing an IT risk awareness program?

   A. Why technology risk is owned by IT
   B. How technology risk can impact each attendee's area of business
   C. How business process owners can transfer technology risk
   D. Why technology risk is more difficult to manage compared to other risk

1-7   It is **MOST** important that risk appetite is aligned with business objectives to ensure that:

    A. resources are directed toward areas of low risk tolerance.
    B. major risk is identified and eliminated.
    C. IT and business goals are aligned.
    D. the risk strategy is adequately communicated.

1-8   A poor choice of passwords and transmission over unprotected communication lines are examples of:

    A. vulnerabilities.
    B. threats.
    C. probabilities.
    D. impacts.

PPT = people Process & Technology

Governance has several goal

C. verifies that organisation.

Audit Charter

Rules and ongagenter

ANSWERS TO SELF-ASSESSMENT QUESTIONS

Correct answers are shown in **bold**.

1-1 A. Effectiveness deals with information being relevant and pertinent to the business process as well as being delivered in a timely, correct, consistent and usable manner. While the lack of system resilience can in some cases affect effectiveness, resilience is more closely linked to the business information requirement of availability.

   B. Confidentiality deals with the protection of sensitive information from unauthorized disclosure. While the lack of system resilience can in some cases affect data confidentiality, resilience is more closely linked to the business information requirement of availability.

   C. Integrity relates to the accuracy and completeness of information as well as to its validity in accordance with business values and expectations. While the lack of system resilience can in some cases affect data integrity, resilience is more closely linked to the business information requirement of availability.

   D. Availability relates to information being available when required by the business process—now and in the future. Resilience is the ability to provide and maintain an acceptable level of service during disasters or when facing operational challenges.

1-2 A. In the risk response planning process, appropriate responses are chosen, agreed on and included in the risk register.

   B. Risk monitoring and control often requires identification of new risk and reassessment of risk. Outcomes of risk reassessments, risk audits and periodic risk reviews trigger updates to the risk register.

   C. While the risk register details all identified risk, including description, category, cause, probability of occurring, impact(s) on objectives, proposed responses, owners and current status, the primary outputs from risk identification are the initial entries into the risk register.

   D. Risk management strategy planning describes how risk management will be structured and performed.

1-3 **A. Assessing how the law may affect the enterprise is the best course of action. The analysis must also determine whether existing controls already address the new requirements.**

   B. Ensuring that necessary adjustments are implemented during the next review cycle is not the best answer, particularly when the law does affect the enterprise. While an annual review cycle may be sufficient in general, significant changes in the internal or external environment should trigger an *ad hoc* reassessment.

   C. Initiating an *ad hoc* amendment to the corporate policy may be a rash and unnecessary action.

   D. Notifying the system custodian to implement changes is inappropriate. Changes to the system should be implemented only after approval by the process owner.

1-4 A. Business objectives are established by the directors as a function of overall organizational governance.

   B. The business strategy is developed by top executives to meet the business objectives set out by the directors.

   C. The risk management strategy is developed by risk managers on the basis of the framework established by the risk governance function.

   D. Risk governance is a strategic business function that helps ensure the risk management strategy is aligned with the overall business strategy.

1-5   A.   **Having an interdisciplinary team contribute to risk management ensures that all areas are adequately considered and included in the risk assessment processes to support an enterprisewide view of risk.**

     B.   Engaging a third party to perform a risk assessment may provide additional expertise to conduct the risk assessment; but without internal knowledge, it will be difficult to assess the adequacy of the risk assessment performed.

     C.   A risk assessment performed by the enterprise's IT department is unlikely to reflect the view of the entire enterprise.

     D.   The internal compliance department ensures the implementation of risk responses based on the requirement of management. It generally does not take an active part in implementing risk responses for items that do not have regulatory implications.

1-6   A.   IT does not own technology risk. An appropriate topic of IT risk awareness training may be the fact that many types of IT risk are owned by the business. One example may be the risk of employees exploiting insufficient segregation of duties (SoD) within an enterprise resource planning (ERP) system.

     B.   **Stakeholders must understand how the IT-related risk impacts the overall business.**

     C.   Transferring risk is not of primary consideration in developing a risk awareness program. It is a part of the risk response process.

     D.   Technology risk may or may not be more difficult to manage than other types of risk. Although this is important from an awareness point of view, it is not as primary as understanding the impact in the area of business.

1-7   A.   **Risk appetite is the amount of risk that an enterprise is willing to take on in pursuit of value. Aligning it with business objectives allows an enterprise to evaluate and deploy valuable resources toward those objectives where the risk tolerance (for loss) is low.**

     B.   There is no link between aligning risk appetite with business objectives and identification and elimination of major risk. Moreover, risk cannot be eliminated; it can be reduced to an acceptable level using various risk response options.

     C.   Alignment of risk appetite with business objectives does converge IT and business goals to a point, but alignment is not limited to these two areas. Other areas include organizational, strategic and financial objectives, among other objectives.

     D.   Communication of the risk strategy does not depend on aligning risk appetite with business objectives.

1-8   A.   **Vulnerabilities represent characteristics of information resources that may be exploited by a threat.**

     B.   Threats are circumstances or events with the potential to cause harm to information resources.

     C.   Probabilities represent the likelihood of the occurrence of a threat.

     D.   Impacts represent the outcome or result of a threat exploiting a vulnerability.

NOTE: For more self-assessment questions, you may also want to obtain a copy of the *CRISC™ Review Questions, Answers & Explanations Manual 2015*, which consists of 400 multiple-choice study questions, answers and explanations, and the *CRISC™ Review Questions, Answers & Explanations Manual 2015 Supplement*, which consists of 100 new multiple-choice study questions, answers and explanations.

SUGGESTED RESOURCES FOR FURTHER STUDY

In addition to the resources cited throughout this manual, the following resources are suggested for further study in this domain (publications in **bold** are stocked in the ISACA Bookstore):

Abkowitz, Mark D.; *Operational Risk Management: A Case Study Approach to Effective Planning and Response*, John Wiley & Sons, USA, 2008

The Association of Insurance and Risk Managers (Airmic); The Public Risk Management Association (Alarm); The Institute of Risk Management (IRM); *A Structured Approach to Enterprise Risk Management (ERM) and the Requirements of ISO 31000*, United Kingdom, 2010

Fraser, John; Betty J. Simkins (Eds.); *Enterprise Risk Management: Today's Leading Research and Best Practices for Tomorrow's Executives*, John Wiley & Sons, USA, 2010

Hillson, David; Ruth Murray-Webster; *Understanding and Managing Risk Attitude*, Gower Publishing Company, United Kingdom, 2005

The Institute of Internal Auditors—Australia; *HB 158-2010 Delivering assurance based on ISO 31000:2009—Risk management - Principles and guidelines*, Standards Australia GPO, 2010, Australia

The Institute of Risk Management (IRM); *Risk Appetite and Tolerance*, United Kingdom, 2011

International Organization for Standardization/International Electrotechnical Commission; *ISO/IEC 27001:2013 Information Technology—Security Techniques—Information Security Management Systems—Requirements*, Switzerland, 2013

ISO/IEC; *ISO/IEC 27005:2011—Information technology—Security techniques—Information security risk management*, Switzerland, 2011

International Organization for Standardization; *ISO 31000:2009 Risk management—Principles and guidelines*, Switzerland, 2009

ISACA, *COBIT® 5 for Risk*, USA, 2013, *www.isaca.org/cobit*

Moeller, Robert, R.; *COSO Enterprise Risk Management: Establishing Effective Governance, Risk, and Compliance Processes, 2nd Edition*, John Wiley & Sons, USA, 2011

National Institute of Standards and Technology (NIST); *NIST Special Publication 800-30 Revision 1: Guide for Conducting Risk Assessments*, USA, 2012

NIST; *NIST Special Publication 800-39: Managing Information Security Risk*, USA, 2011

Pickett, K. H. Spencer; *Auditing the Risk Management Process*, John Wiley & Sons, USA, 2013

The Risk Management Society (RIMS); The Institute of Internal Auditors (IIA); *Risk Management and Internal Audit: Forging a Collaborative Alliance*, USA, 2012

Rittenberg, Larry; Frank Martens; *Enterprise Risk Management: Understanding and Communicating Risk Appetite*, Committee of Sponsoring Organizations of the Treadway Commission (COSO), USA, 2012

Sobel, Paul J.; *Auditor's Risk Management Guide: Integrating Auditing and ERM*, Wolters Kluwer/CCH, USA, 2013

Walker, Paul L.; William G. Shenkir; Thomas L. Barton; *Improving Board Risk Oversight Through Best Practices*, The Institute of Internal Auditors Research Foundation, USA, 2011

Section Two: Content

1.0 OVERVIEW

The risk management process includes identifying, assessing, mitigating and monitoring risk. As seen in **exhibit 1.1**, this is a cyclical process that builds on the information gathered in the previous phase.

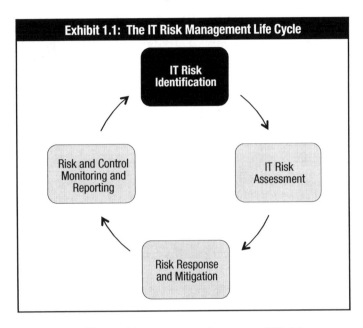

The output of IT risk identification facilitates risk assessment, the output of IT risk assessment supports the risk response decision, and the output of risk response indicates the controls and residual risk that must be monitored. The risk monitoring phase leads to the refinement of the risk response and the reinitiation of risk identification and assessment. IT risk management follows these phases with a focus on the risk related to not only the IT department, but all IT-enabled risk to an organization.

1.1 IT RISK MANAGEMENT GOOD PRACTICES

The process of IT risk management should follow a structured methodology based on good practices and a desire to seek continuous improvement. When starting a risk management effort, the risk practitioner should review the current risk management practices of the organization in relation to the processes of risk identification, assessment, response and monitoring. This will determine whether the organization's IT risk management program is based on acceptable and recognized good practices.

Using good practices can assist in the development of a consistent enterprisewide risk management program. IT risk management practices may be based on an international standard or on another risk management model to ensure that the risk management program is complete and authoritative. Examples of good practices include:
• *COBIT® 5 for Risk*
• Committee of Sponsoring Organizations of the Treadway Commission (COSO)
• *International Organization for Standardization (ISO)/International Electrotechnical Commission (IEC) 27005:2011*
  —Information technology—Security techniques—Information security risk management
• *ISO 31000:2009 – Risk Management Principles and Guidelines*
• *National Institute of Standards and Technology (NIST) Special Publication 800-30 Revision 1: Guide for*
  Conducting Risk Assessments
• *NIST Special Publication 800-39: Managing Information Security Risk*

An example of a recognized risk management program based on ISO/IEC 27005 includes several components, as shown in **exhibit 1.2**.

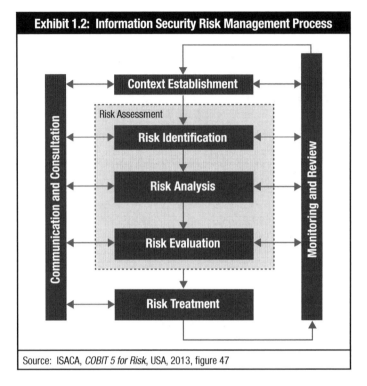

Exhibit 1.2: Information Security Risk Management Process

Source: ISACA, *COBIT 5 for Risk*, USA, 2013, figure 47

Exhibit 1.3 provides a summary of the important concepts of these process steps.

Exhibit 1.3: ISO/IEC 27005 Process Steps	
ISO/IEC 27005 Process Step	**Important Concepts of the Component**
Context Establishment	This process step includes: • Setting the basic criteria necessary for establishment of information security risk management (ISRM) • Defining the scope and boundaries • Establishing an appropriate organization operating the ISRM
Risk Assessment	Risk assessment determines the value of the information assets, identifies the applicable threats and vulnerabilities that exist (or could exist), identifies the existing controls and their effect on the risk identified, determines the potential consequences and finally, prioritizes the derived risk and ranks it against the risk evaluation criteria set in the context establishment. This process step consists of risk identification, risk analysis and risk evaluation.
Risk Identification	Risk identification includes the identification of: • Assets • Threats • Vulnerabilities • Existing controls • Consequences The output of this process is a list of incident scenarios with their consequences related to assets and business processes.
Risk Analysis	The risk estimation step includes: • Assessment of consequences • Assessment of incident likelihoods • Determination of level of risk
Risk Evaluation	In this step, levels of risk are compared according to risk evaluation criteria and risk acceptance criteria. The output is a prioritized list of risk elements and the incident scenarios that lead to the identified risk elements.

Chapter 1: IT Risk Identification — Section Two: Content

CRISC
Certified in Risk
and Information
Systems Control™
An ISACA® Certification

Exhibit 1.3: ISO/IEC 27005 Process Steps *(cont.)*	
ISO/IEC 27005 Process Step	**Important Concepts of the Component**
Risk Treatment	Risk treatment options include: • Risk modification • Risk retention • Risk avoidance • Risk sharing
Risk Acceptance	The input is a risk treatment plan and the residual risk assessment subject to the risk acceptance criteria. This stage comprises the formal acceptance and recording of the suggested risk treatment plans and residual risk assessment by management, with justification for those that do not meet the enterprise's criteria.
Risk Communication and Consultation	This is a transversal process where information about risk should be exchanged and shared between the decision maker and other stakeholders through all the steps of the risk management process.
Risk Monitoring and Review	Risk and its influencing factors should be monitored and reviewed to identify any changes in the context of the organization at an early stage and to maintain an overview of the complete risk picture.

The IT risk management program should be:
• Comprehensive (thorough, detailed)
• Complete (carried through to the end)
• Auditable (reviewable by an independent third party)
• Justifiable (based on sound reasoning)
• Legal (in compliance with regulations)
• Monitored (subject to review and accountability)
• Enforced (consistent, mandated and required)
• Up to date (current with changing business processes, technologies and laws)
• Managed (adequately resourced, with oversight and support)

Good practices in IT risk management are, like all other good practices, subject to change and improvement. The risk practitioner should be aware of the changes in good practices to ensure that the current risk practices of the organization follow recognized good practices and that the risk response of the organization is justifiable and in line with current standards and practices.

NOTE: The CRISC candidate will not be tested on any specific standard. The use of any standards in this review manual is for example and explanatory purposes only.

1.1.1 Risk Identification and Classification Standards and Frameworks
Several good sources for risk identification and classification standards and frameworks are available to the risk practitioner. The following list is not comprehensive, and many more standards are available. However, this list may allow the risk practitioner to consider a framework or standard that would be suitable for use in his/her organization. Many countries and industries have specific standards that must be used by organizations operating in their jurisdiction. The use of a recognized standard may provide credibility and completeness for the risk assessment and management program of the organization and help ensure that the risk management program is comprehensive and thorough.

ISO 31000:2009 Risk Management—Principles and Guidelines
ISO 31000:2009 states:

This international standard recommends that organizations develop, implement and continuously improve a framework whose purpose is to integrate the process for managing risk into the organization's overall governance, strategy and planning, management, reporting processes, projects and activities.

Although the practice of risk management has been developed over time and within many sectors in order to meet diverse needs, the adoption of consistent processes within a comprehensive framework can help to ensure that risk is managed effectively, efficiently and coherently across an organization. The generic approach described in this standard provides the principles and guidelines for managing any form of risk in a systematic, transparent and credible manner and within any scope and context.[1]

COBIT® 5 for Risk

COBIT 5 for Risk is described as follows:

COBIT 5 provides a comprehensive framework that assists enterprises in achieving their objectives for the governance and management of enterprise information technology (IT). Simply stated, COBIT 5 helps enterprises to create optimal value from IT by maintaining a balance between realising benefits and optimising risk levels and resource use. COBIT 5 enables IT to be governed and managed in a holistic manner for the entire enterprise, taking into account the full end-to-end business and IT functional areas of responsibility and considering the IT-related interests of internal and external stakeholders.

COBIT 5 for Risk ... *builds on the COBIT 5 framework by focusing on risk and providing more detailed and practical guidance for risk professionals and other interested parties at all levels of the enterprise.[2]*

IEC 31010:2009 Risk Management—Risk Assessment Techniques

IEC 31010:2009 states:

Organizations of all types and sizes face a range of risks that may affect the achievement of their objectives.

These objectives may relate to a range of the organization's activities, from strategic initiatives to its operations, processes and projects, and be reflected in terms of societal environmental, technological, safety and security outcomes, commercial, financial and economic measures, as well as social, cultural, political and reputation impacts.

All activities of an organization involve risks that should be managed. The risk management process aids decision making by taking account of uncertainty and the possibility of future events or circumstances (intended or unintended) and their effects on agreed objectives.[3]

ISO/IEC 27001:2013 Information Technology—Security Techniques—Information Security Management Systems – Requirements

ISO 27001:2013 states:

The organization shall define and apply an information security risk assessment process that identifies the information security risks: 1) apply the information security risk assessment process to identify risks associated with the loss of confidentiality, integrity and availability for information within the scope of the information security management system and 2) identify risk owners.[4]

ISO/IEC 27005:2011 Information Technology—Security Techniques—Information Security Risk Management

ISO/IEC 27005 states:

This international standard provides guidelines for information security risk management in an organization, supporting in particular the requirements of an information security management system (ISMS) according to ISO/IEC 27001. However, this standard does not provide any specific methodology for information security risk management. It is up to the organization to define their approach to risk management, depending for example on the scope of the ISMS, context of risk management, or industry sector. A number of existing methodologies can be used under the framework described in this standard to implement the requirements of an ISMS.[5]

NIST Special Publications

NIST has a wide range of special publications available at csrc.nist.gov. Some of the publications related to IT risk follow.

NIST Special Publication 800-30 Revision 1: Guide for Conducting Risk Assessments

NIST Special Publication 800-30 Revision 1 describes risk assessment in the following manner:

- *Risk assessments are a key part of effective risk management and facilitate decision making at all three tiers in the risk management hierarchy including the organization level, mission/business process level, and information system level.*

• *Because risk management is ongoing, risk assessments are conducted throughout the system development life cycle, from pre-system acquisition (i.e., material solution analysis and technology development), through system acquisition (i.e., engineering/manufacturing development and production/deployment), and on into sustainment (i.e., operations/support).*[6]

NIST Special Publication 800-39: Managing Information Security Risk

NIST Special Publication 800-39 states:

The purpose of Special Publication 800-39 is to provide guidance for an integrated, organization-wide program for managing information security risk to organizational operations (i.e., mission, functions, image, and reputation), organizational assets, individuals, other organizations, and the Nation resulting from the operation and use of federal information systems. Special Publication 800-39 provides a structured, yet flexible approach for managing risk that is intentionally broad-based, with the specific details of assessing, responding to, and monitoring risk on an ongoing basis provided by other supporting NIST security standards and guidelines.[7]

1.2 METHODS TO IDENTIFY RISK

Risk identification is the process for discovering, recognizing and documenting the risk an organization faces.

Some ways to identify risk include:
• Historical- or evidence-based methods, such as review of historical events, for example, the use of checklists and the reviews of past issues or compromise
• Systematic approaches (expert opinion), where a risk team examines and questions a business process in a systematic manner to determine the potential points of failure
• Inductive methods (theoretical analysis), where a team examines a process to determine the possible point of attack or compromise

There are several types of risk that the risk practitioner should consider in risk identification, shown in **exhibit 1.4**.

Exhibit 1.4: Business-related IT Risk Types	
Type	**Description**
Investment or expense risk	The risk that the IT investment fails to provide value for money or is otherwise excessive or wasteful; this includes consideration of the overall IT investment portfolio
Access or security risk	The risk that confidential or otherwise sensitive information may be divulged or made available to those without appropriate authority; an example of this risk is noncompliance with local, national and international laws related to privacy and protection of personal information
Integrity risk	The risk that data cannot be relied on because they are unauthorized, incomplete or inaccurate
Relevance risk	The risk associated with not getting the right information to the right people (or process or systems) at the right time to allow the right action to be taken
Availability risk	The risk of loss of service or the risk that data are not available when needed
Infrastructure risk	The risk that an enterprise does not have an IT infrastructure and systems that can effectively support the current and future needs of the business in an efficient, cost-effective and well-controlled fashion (includes hardware, networks, software, people and processes)
Project ownership risk	The risk of IT projects failing to meet objectives through lack of accountability and commitment

ISO/IEC 27005 and *NIST Special Publication 800-30 Revision 1: Guide to Conducting Risk Assessment* both describe the process of risk identification in a similar series of steps, as seen in **exhibit 1.5**. The results of the risk identification process will then feed into the risk estimation and assessment process.

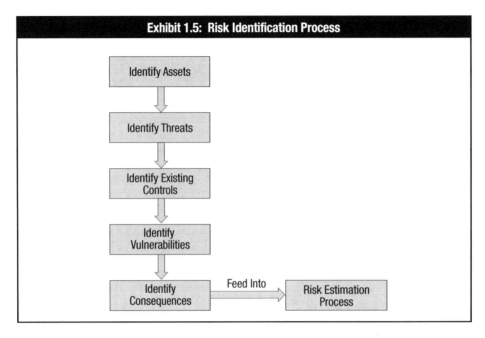

Each risk should be documented as it is identified in a risk register to permit the tracking of the risk in a common reference location.

Many sources of risk documentation can assist the risk practitioner in identifying the threats and risk faced by the organization, including:
• Audit reports
• Incident reports
• Public media (e.g., newspapers, television, etc.)
• Annual reports
• Press releases
• Vulnerability assessments and penetration tests
• Business continuity and disaster recovery plans
• Interviews and workshops with managers, employees, customers, suppliers and auditors
• Threat intelligence services

1.2.1 Conducting Interviews
Gathering information from staff includes challenges that the risk practitioner must be aware of, including:
• Exaggeration: Everyone wants their department to be seen as critical and essential.
• Inaccuracies: People may not correctly understand the overall business process or dependencies between departments.

Good interviewing practices include:
• Designate a specific time period and do not exceed that time unless it is mutually agreed that the interview continue. When a manager is told that a staff member will be needed for 45 minutes, he/she should not discover that the interview lasted 90 minutes.
• Know as much as possible about the business process in advance of the interview. Doing this can help avoid spending time on explanations and trying to gain an understanding of core business functions.
  – Obtain and review process documentation (i.e., business process maps, standard operating procedures, business impact analysis [BIA] results, network topologies)

- Prepare questions and provide them to the interviewee in advance so that any supporting documentation, reports or data are available at the time of the interview.
- Conduct interviews with senior leadership including board members, managers, users, administrators, critical third-party service providers, customers and suppliers to ensure a thorough understanding of the enterprise, including every aspect of each business operation. Encourage interviewees to be open about challenges they face and risk that concerns them. Encourage them also to describe any potential missed opportunities or problems associated with their current processes, systems and services/products. They know the systems best.

An important detail to obtain during the interview is the level of impact that previous incidents have had on the organization including how the incident was handled, results of post-incident review and root cause analysis, and current status of any noted remediation activities from prior incidents. Also, it is important to learn the way that the incident was handled, whether a follow-up was done to learn from the incident, and whether the recommendations from the previous incident were addressed and resolved.

Note that the people available for interviews may not be the correct people. Management may not want key personnel to be taken away from important work for the interview and may only provide access to personnel who are not the best sources of information.

1.3 RISK CULTURE AND COMMUNICATION

Risk management is a core part of corporate governance. The governance of the assets and mission of the organization is reflected in the ways in which the organization seeks to protect its assets and attain its goals. Risk is a factor that may lead to failure or a loss of asset value. Understanding risk is understanding the business's goals, objectives, values and ethics. Senior management either consciously or unconsciously develops an attitude toward risk that indicates their willingness to embrace, cautiously accept or avoid risk. This is called the risk culture. The best indicator of the organization's risk culture is how the organization handles risk.

1.3.1 Risk Culture

Risk culture reflects a balance between weighing the negative, positive and regulatory elements of risk, as seen in **exhibits 1.6** and **1.7**. Symptoms of an inadequate or problematic risk culture are listed in **exhibit 1.8**.

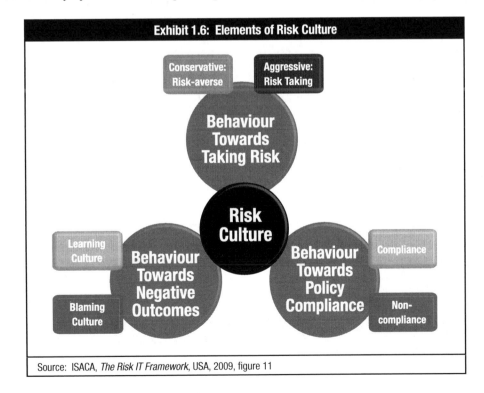

Exhibit 1.6: Elements of Risk Culture

Source: ISACA, *The Risk IT Framework*, USA, 2009, figure 11

Exhibit 1.7: Description of Risk Culture Elements	
Element	**Description**
Behavior toward taking risk	How much risk does the enterprise feel it can absorb, and what specific risk is it willing to take?
Behavior toward policy compliance	To what extent will people embrace and/or comply with policy?
Behavior toward negative outcomes	How does the enterprise deal with negative outcomes, i.e., loss events or missed opportunities? Will it learn from them and try to adjust, or will blame be assigned without treating the root cause?

Exhibit 1.8: Symptoms of an Inadequate or Problematic Risk Culture	
Symptom	**Description**
Misalignment between real risk appetite and translation into policies	Management's real position toward risk can be reasonably aggressive and risk taking, whereas the policies that are created reflect a much stricter attitude.
Existence of a "blame culture"	This type of culture should, by all means, be avoided; it is the most effective inhibitor of relevant and efficient communication. In a blame culture, business units tend to point the finger at IT when projects are not delivered on time or do not meet expectations. In doing so, they fail to realize how the business unit's involvement upfront affects project success. In extreme cases, the business unit may assign blame for a failure to meet the expectations that the unit never clearly communicated. The "blame game" only detracts from effective communication across units, further fueling delays. Executive leadership must identify and quickly control a blame culture if collaboration is to be fostered throughout the enterprise.

1.3.2 Risk Communication

The method and openness of risk communications also play a key role in defining and understanding the risk culture of the organization. Risk communication removes the uncertainty and doubts concerning risk management. If risk is to be managed and mitigated, it must first be discussed and effectively communicated in an appropriate level to the various stakeholders and personnel throughout the organization.

The benefits of open communication on risk include:
• Assistance in executive management's understanding of the actual exposure to risk, enabling the definition of appropriate and informed risk responses
• Awareness among all internal stakeholders of the importance of integrating risk management into their daily duties
• Transparency to external stakeholders regarding the actual level of risk and risk management processes in use

The consequences of poor communication on risk include:
• A false sense of confidence at all levels of the enterprise and a higher risk of a breach or incident that could have been prevented. Risk ignorance is an unacceptable risk management strategy.
• Lack of direction or strategic planning to mandate risk management efforts
• Unbalanced communication to the external world on risk, especially in cases of high, but managed, risk, which may lead to an incorrect perception on actual risk by third parties such as:
  – Clients
  – Investors
  – Regulators
• The perception that the enterprise is trying to cover up known risk from stakeholders

Exhibits 1.9 and **1.10** describe the broad array of information flows and the major types of IT risk information that should be communicated.

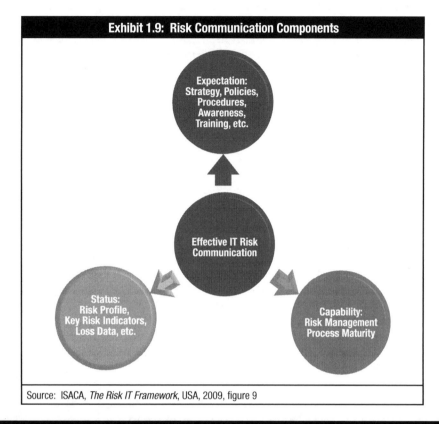

Exhibit 1.9: Risk Communication Components

Expectation:
Strategy, Policies, Procedures, Awareness, Training, etc.

Effective IT Risk Communication

Status:
Risk Profile, Key Risk Indicators, Loss Data, etc.

Capability:
Risk Management Process Maturity

Source: ISACA, *The Risk IT Framework*, USA, 2009, figure 9

| Exhibit 1.10: Risk Components to Be Communicated ||
Type	Description
Expectations from risk management	The risk components that must be communicated throughout the enterprise include risk strategy, policies, procedures, awareness training and continuous reinforcement of principles. This is essential communication regarding the enterprise's overall strategy toward IT risk and: • Drives all subsequent efforts on risk management • Sets the overall expectations about the risk management program
Current risk management capability	This information: • Allows for monitoring of the state of the "risk management engine" in the enterprise • Is a key indicator for good risk management • Has predictive value for how well the enterprise is managing risk and reducing exposure
Status	This includes the actual status with regard to IT risk, including information such as: • The risk profile of the enterprise, i.e., the overall portfolio of (identified) risk to which the enterprise is exposed • Key risk indicators (KRIs) to support management reporting on risk • Event/loss data • The root cause of loss events • Options to mitigate risk (including cost and benefits)

The Value of Communication

The discipline of risk management is related to audit, business continuity management and security, and having a reporting relationship that facilitates communication among those groups is desirable. The ability to share data regarding risk, incidents, vulnerabilities and assets can result in a greater level of accuracy for the risk management process.

Communication of risk incidents is an essential part of the risk management process. While all risk incidents do not need to be reported, a lack of communication of risk events can be a sign that an organization is not healthy or stable. If management levels are not open and transparent with regard to issues and operational failures, decisions may be made based on incomplete or inaccurate information or management may not be aware or understand the current situation in other departments or management levels. The risk practitioner should seek to develop lines of communication and reporting so that information is available to management on a timely basis and encourage communication of even negative activities when appropriate.

1.4 THE BUSINESS'S IT RISK STRATEGY

Understanding the business's risk strategy helps to develop an IT risk strategy. IT risk is the IT-enabled business risk that stems from the use of IT. All IT risk must be measured not only by its impact on IT services but by the impact of risk on business operations.

Most organizations have some form of risk management in place, even if it is not formally defined or is somewhat immature. The risk practitioner must investigate the business's current environment and confirm that documentation of controls, risk, audit and regulations that impact the determination of the level of risk is available.

1.4.1 Buy-in of Senior Management

A continuing thread through the risk management process is the importance of senior management support. Without the active support of senior management, the risk management process is almost always unsuccessful. With the support of senior management, the risk management process is much more likely to have the budget, authority, access to personnel and information, and legitimacy that will provide a successful result. Senior management support should be visible and active, and they should be willing to intervene when necessary to communicate the importance of risk identification efforts and the need for everyone to actively contribute at the request of the risk practitioner.

1.4.2 Alignment With Business Goals and Objectives

Risk management depends on business goals and objectives. The risk practitioner must be careful not to consider risk from the perspective of one department or one process without also considering the risk to other departments, business partners or the overall business goals. It can be easy, especially for IT risk practitioners, to only calculate risk based on the impact on IT and forget to also calculate the impact of a risk on the business processes that the IT systems support. Therefore, a risk practitioner must look beyond his/her department.

The best way to understand the goals and objectives of the organization is to communicate with senior management. The challenge is that, in some cases, the information they can provide may be censored due to ongoing negotiations or confidential strategic plans. The risk practitioner should consult with senior management to gain an understanding of their vision and strategy. This includes discussing lines of business, new technologies, future growth or changes in priorities.

> Remember that some lines of business that are currently prominent may not be a part of the future strategy of the organization; the senior management team may already be planning to move away from one product line and toward a new emerging product.

Risk considers future strategy and developments and the risk associated with pursuing or not pursuing those ventures. The risk associated with a new venture may be related financial loss, reputational damage, changes in employment agreements and different regulatory environments.

The underlying importance of risk management in relation to business goals and strategy is to ensure that the risk is closely aligned with, and integrated into, the strategy, vision and direction of the organization. Remember that the organization will choose a path that seems best for its success regardless of the resistance of risk management. The risk practitioner should seek to:
• Understand the business
• Listen to the strategy
• Proactively seek out ways to secure new technologies and business processes

- Build relationships and communication infrastructure to weave risk management into each business process and new project
- Be aware of and mitigate the risk of change
- Work to create a culture that encourages the participation of risk management into business processes

A critical component of risk management is history. The events of the past must be considered in the light of the risk management effort. An organization that has not taken steps to mitigate the future impact of events that have already occurred can seem irresponsible. However, a narrow focus on the past may result in a lack of watching for new threats and future issues. Therefore, the organization may fall into the habit of only fixing events once they have occurred. Another consideration is that an organization was fortunate in the past, and an event that could have had a devastating impact only had a minor impact, which was quickly resolved. This can lead to overconfidence, which can be dangerous and cause the organization to overlook serious problems that should have been thoroughly addressed and mitigated. In this case, the preexisting conditions that led to the event may remain and could cause a much more serious incident.

When an organization goes through a merger or acquisition, new risk emerges. A merger creates uncertainty and stress that can result in poor judgment or inappropriate actions by personnel. Many personnel may be worried about protecting their jobs, being moved into other positions, having to compete for a new assignment or the company's stability. In many cases, a merger will result in job loss, and some studies have suggested that employees who are about to lose their jobs are likely to take corporate assets and information with them in more than half of the cases.

Downsizing is also a risk because many employees have assets of the organization in their possession, including access to sensitive or protected information. When a downsizing initiative is being planned or announced, the risk manager should actively increase the controls over sensitive information, be increasingly diligent to enforce monitoring and work with human resources (HR) to identify potentially risk-laden scenarios.

In the end, if IT risk management does not move in the same direction as the business, then the risk management function will quickly become distant from the business processes, irrelevant and ineffective.

1.4.3 Organizational Structures and Impact on Risk

The effectiveness of the risk management effort is often influenced by the positioning of the risk management function within the organizational structure. Ideally, risk management should be an enterprisewide function with the ability to reach into all the parts of the organization and provide leadership, advice and direction on risk management. However, this is not always the case. Risk management is often only one of the job responsibilities of a manager or it is a small function buried deep within one area of the business. An effective risk management program provides a consistent way to manage risk and creates a risk framework that serves as a foundation for risk management for all departments and business functions. Ideally, the organization will have established three lines of defense, with risk being managed by the front line and guided, directed, influenced and/or assessed by the second line, and with independent oversight, review and monitoring by the third line. In reality, these lines are often blurred.

Risk management may be applied to an entire organization as one or more formal risk management teams, or it may be practiced separately in each level of the organization or in regard to specific functions, projects and activities. However, it can be unclear who the risk practitioner should report to within the organization. Many companies do not have a dedicated chief risk officer (CRO) who reports to the chief executive officer (CEO) or board of directors.

The size and diversity of the organization is a key factor in managing risk. In a small company operating in one primary market, the management of risk can be fairly straightforward. However, even in a small company, the cultural differences between departments can be extreme. The approach to risk in the finance department may be diametrically opposed to the approach to risk in the sales department. This requires the risk practitioner to work with the management team to develop an understanding of risk and a standard approach to measuring and mitigating risk.

In a large organization, risk management may be too large of a task for one department or team. If the organization operates in several countries or supports a wide variety of products or services, the risk management effort may need to be arranged by department, products, services or geographic region. In most cases, the organization of the risk management group should follow the same model and the organization of the business continuity management team.

RACI (Responsible, Accountable, Consulted, Informed)

There are four main types of roles that are involved in the risk management process:
- The individuals responsible for managing the risk
- The individuals accountable for the risk management effort
- The individuals who provide support and assistance to the risk management effort (consulted)
- The individuals who evaluate or monitor the effectiveness of the risk management effort (informed)

The use of a RACI model can assist in outlining the roles and responsibilities of the various stakeholders. The purpose of a RACI model is to clearly show the relationships between the various stakeholders, the interaction between the stakeholders and the roles that each stakeholder plays in the successful completion of the risk management effort. **Exhibit 1.11** describes the components of a RACI model. An example of a RACI chart is shown in **exhibit 1.12**.

Exhibit 1.11: RACI Model	
Assigned Role and Responsibility	**Description**
Responsible	The person(s) tasked with getting the job done. This is the role of the person(s) performing the actual work effort to meet a stated objective.
Accountable	The person accountable (liable, answerable) for the completion of the task. He/she is responsible for the oversight and management of the person(s) responsible for performing the work effort. He/she may also play a role in the project and bear the responsibility for project success or failure. Accountability should be with a sole role or person in order to be effective.
Consulted	The person(s) consulted as a part of the project. They may provide input data, advice, feedback or approvals. Consulted personnel may be from other departments, from all layers of the organization, from external sources or from regulators.
Informed	The person(s) who are informed of the status, achievement and/or deliverables of the task, but who are often not directly responsible for the work effort.

Exhibit 1.12: Sample RACI Chart				
Task	**Senior Management**	**Steering Committee (Chair)**	**Department Managers**	**Risk Practitioner**
Collect risk data	I	A	C	R
Deliver the risk report	I	A	I	R
Prioritize risk response	A	I	R	C
Monitor risk	I	A	R	C

1.4.4 Organizational Culture, Ethics and Behavior and the Impact on Risk

One of the first challenges faced by a risk practitioner is to determine the risk acceptance levels of senior management. An additional challenge is that even after the risk appetite is known, senior management may change, requiring the determination of the new approach by the new managers. This is complicated by risk being such a variable calculation. Depending on market conditions, confidence, past successes or failures, global economics, reports in the media, availability of resources, new regulations, or long-term strategy, the risk appetite of senior management may change dramatically. Examples of risk that management must evaluate and accept include deciding whether to:
- Invest
- Take on a new line of business
- Develop a new product
- Open a new office
- Hire a new employee
- Invest in new hardware or software
- Upgrade existing applications
- Implement new controls

Culture

All employees of the organization should be aware of the risk culture. A culture drives the behaviors of personnel, and people will often act according to their environment. Risk culture is defined as the set of shared values and beliefs that governs attitudes toward risk-taking, care and integrity and determines how openly risk and losses are reported and discussed.

A culture of honesty and openness may reduce the risk of theft, inappropriate actions or attacks. In many cases, the way to influence the behavior of people is to influence their ethics and beliefs. A person will often act according to their beliefs, so creating an environment that addresses people's belief systems is often an effective method of changing behaviors as well.

An organization could have a subculture in a department that is different from the organizational culture. This may lead to people taking more risk than management thinks is wise or to people refusing to take risk when management would like them to do so. Management may encourage employees to take risk, but then blame them when something is wrong, thus discouraging creativity or innovation.

Ethics

Risk is often impacted by the ethics of the personnel of the organization. It is easy to understand that an organization with poor ethical standards may be more susceptible to fraud or theft, but this risk may also apply to an organization that has poor management processes in place to identify errors, misuse or fraud. Ethics are related to an individual's perception of right and wrong and are not necessarily linked to the law. There are many things that may not be illegal, but may still be seen as unethical. For example, in many organizations, it is acceptable to receive gifts from clients or suppliers, whereas in others it is not acceptable. To address the risk of a person violating the policy of the organization, the policy must be communicated to everyone and visibly enforced.

Ethics also applies to how people believe that they have been treated. An employee who feels that he/she has been poorly treated may seek revenge, and this can lead to a serious risk. A well-treated employee may be an active supporter of proper behavior, and this will reduce risk.

1.4.5 Compliance With Laws, Regulations, Standards and Compliance Requirements

Organizations are required to comply with the laws and regulations of the jurisdictions in which they operate. Therefore, it is important to know what laws apply and to understand the requirements of these laws. This can be challenging because many laws are open to interpretation and required levels of compliance are not always stipulated. For example, many laws require adequate protection of sensitive data without specifying what would be an adequate level of protection. Laws and regulations often contradict one another, and an organization that operates in various regions of the world may find it difficult to comply with the all of the laws in each location.

Organizations that operate globally or even within different regions of one country may build a global program of policies and a control suite to handle the common regulations and then have a regional or nation-specific addendum to handle the exceptions and their controls. In some cases an organization may choose not to be compliant with certain regulations if the cost of compliance is greater than the fine or consequences associated with noncompliance.

Regulations often require the organization to report on whether they are compliant with the regulations. Failure to comply with regulations may result in financial penalties or loss of a license to operate as well as damage to the reputation of the organization. An organization should keep logs and records that can be used to validate whether information or an information system has been accessed improperly.

An example of a standard that can be used to measure improper access is the Payment Card Industry Data Security Standard (PCI DSS). This standard, created by members of the payment card industry, should be followed by companies that handle payment cards, but it is not required by law. These standards are described in **exhibit 1.13**.

Exhibit 1.13: PCI Data Security Standard—High Level Overview	
The 12 PCI DSS Requirements	
Build and Maintain a Secure Network and Systems	**1.** Install and maintain a firewall configuration to protect cardholder data **2.** Do not use vendor-supplied defaults for system passwords and other security parameters
Protect Cardholder Data	**3.** Protect stored cardholder data **4.** Encrypt transmission of cardholder data across open, public networks
Maintain a Vulnerability Management Program	**5.** Protect all systems against malware and regularly update anti-virus software or programs **6.** Develop and maintain secure systems and applications
Implement Strong Access Control Measures	**7.** Restrict access to cardholder data by business need to know **8.** Identify and authenticate access to system components **9.** Restrict physical access to cardholder data
Regularly Monitor and Test Networks	**10.** Track and monitor all access to network resources and cardholder data **11.** Regularly test security systems and processes
Maintain an Information Security Policy	**12.** Maintain a policy that addresses information security for all personnel
Provided courtesy of PCI Security Standards Council, LLC and/or its licensors. ©2006-2013 PCI Security Standards Council, LLC. All Rights Reserved. *Payment Card Industry (PCI) Data Security Standard*, v3.0	

An example of a law regarding the protection of personal information is the European Union Data Protection Regulation 45/2001:

*These provisions aim to ensure a **high level of protection** for personal data managed by Community institutions and bodies. In particular, such data have to be:*
* *processed fairly and lawfully;*
* *collected for specified, explicit and legitimate purposes and not further processed in a way incompatible with those purposes;*
* *adequate, relevant and not excessive in relation to the purposes for which they are collected and/or further processed;*
* *accurate and, where necessary, kept up to date (all reasonable steps should be taken to ensure that data which are inaccurate or incomplete in relation to the purposes for which they are collected or for which they are further processed, are erased or rectified);*
* *kept in a form which permits identification of data subjects for no longer than is necessary for the purposes for which the data are collected or for which they are further processed[8]*

Reports on compliance should be accurate, complete and submitted in a timely manner. To ensure compliance, an organization must have the ability to monitor and measure the controls in use. Reports should be comparable from one reporting period to another, and any trends or areas of noncompliance should be identified and addressed. In cases where there is noncompliance, a justification for the reasons for noncompliance should be provided.

1.4.6 Establishing an Enterprise Risk Management Approach

Risk management is an enterprisewide activity, and it is usually best to develop a standard, structured approach that can be applied to the entire enterprise. When risk is considered on a system-by-system or project-by-project basis, the result is a spotty risk solution that has many individually good efforts, but no consistency or interoperability among the risk solutions that are implemented. This may mean that risk is measured differently in different areas, and gaps may appear between boundaries of the various projects or systems. A large enterprise consisting of many divisions, departments, lines of business, and products or services may represent many different cultures and business models. That is why the risk management effort should be tailored to the needs of each organization. There is no one risk management approach that is suitable for all types of organizations or even for all components of a large, diverse organization. The risk practitioner must be sensitive to local departmental cultures, priorities, regulations, goals and restraints before recommending a risk management approach or framework.

In some large organizations and governments, the risk management function serves a consultative and advisory role that provides data and recommendations to each department, but does not actually determine the development of the control or risk mitigation framework. In such an organization, risk management is a clearinghouse for information that can be accessed by all departments as a part of systems or project development.

Enterprisewide risk management benefits from clear support from senior management, which should require risk management to be consulted on any new project and ensure the inclusion of recommendations of risk management before signing off on or funding any new projects or business initiatives.

Policy

A critical part of establishing the risk management process is the development and approval of a risk management policy. The policy reflects the attitude and intent of management in relation to risk. This sets the tone for the risk culture of the organization. Policies should include a statement relating to the reasoning or rationale behind the approach to accepting or mitigating risk (e.g., mandating compliance with laws and regulations), setting the accountability for risk (e.g., measuring and reporting on risk), and a commitment to continuous improvement of the risk environment.

1.5 INFORMATION SECURITY RISK CONCEPTS AND PRINCIPLES

IT risk is often linked to information security, which is the protecting of information and information systems (including technology) from risk events. Information security controls are based on risk, and risk is the primary justification used to support information security activities.

Some organizations (such as insurance or financial organizations) have a mature risk management process that is able to quantify risk with a high level of accuracy; however, for many organizations, the term risk may be poorly understood and may represent an emotional value based on perception. Risk is often hard to measure, and because it is based on concepts such as likelihood and impact, it can be hard to quantify. In most cases, the level of impact should an undesired event occur is unpredictable.

Likelihood (also known as probability) is the measure of frequency of which an event may occur. In the area of risk identification, likelihood is used to calculate the level of risk facing an organization based on the number of risk events that may occur within a time period and is often measured on an annual basis.

Other factors that can affect likelihood include:
- **Volatility**: In some cases the probability of a risk varies, depending on the volatility of the situation. When conditions vary greatly, it is harder to predict the likelihood of a given event. It is likely that such a risk would be a higher priority for risk management because of its higher unpredictability. (This is also referred to as dynamic range.)
- **Velocity**: Some analyses include the criterion of risk velocity (or speed of onset). This is a measure of how much prior warning and preparation time an organization may have between the event's occurrence and impact. This, in itself, can be split into speed of reaction and speed of recovery.
- **Proximity**: The time from the event occurring and the impact on the organization is sometimes known as proximity.
- **Interdependency**: It is important not just to consider risk in isolation, but also in various combinations. The materialization of two or more types of risk might impact the organization differently, depending on whether the events occur simultaneously or concurrently.
- **Motivation**: A highly motivated attacker is more likely to persist in an attack and have a higher chance of success.
- **Skill**: A skillful attacker is more likely to be successful than an unskilled attacker.
- **Visibility**: A well-known vulnerability is more likely to be attacked.

Risk impact is the calculation of the amount of loss or damage that an organization may incur due to a risk event. This loss may be measured in quantitative, semiquantitative and qualitative terms.

Risk management should be based on calculated actions and justified controls. This means that a method of determining risk that is not based on emotion or perception must be found. One method of calculating risk is to evaluate the impact of an event on the confidentiality, integrity and availability (CIA) of information and information systems. Remember that the risk associated with an information system is primarily about business risk and the impact that the failure or compromise of a system or information would have on the overall business, not just on IT.

The factors that affect the impact of an event must be measured in two ways: (1) the impact due to a compromise or loss of information and (2) the impact due to the loss or compromise of an information system. The result of the assessment may be different for each of these. In some cases, the impact on the business due to the loss of a system may be more serious than the impact from the loss of information. However, the impact from the compromise of a system is always dependent on the impact due to the compromise of the information contained within the system.

The determination of the impact of a compromise is important to know what controls are necessary to protect the system adequately. The risk justifies the need for controls, and the extent of the controls needed is appropriate according to the associated risk.

CIA is used in several risk management methodologies, often using a qualitative risk assessment approach that will evaluate the impact of a breach according to a range of levels. ISO 27005 uses the levels of very low, low, medium, high and very high. When the risk associated with a data element (information) or an information system is assessed, the risk practitioner, in cooperation with the business stakeholders, must determine the correct level of classification for the breach

1.5.1 Confidentiality

Confidentiality pertains to the requirement to maintain the secrecy and privacy of data. Examples of the need to protect sensitive information are the requirement to identify and protect personally identifiable information (PII), personal health information (PHI) or intellectual property (IP). In many places, some types of data must be protected by law.

A breach of confidentiality means the improper disclosure of information, such as disclosing information to an internal or external resource that was not authorized to access the information. This would be a violation of the principles of "need to know" or "least privilege." There are many factors that could lead to disclosure, including improperly managed access controls, social engineering and aggregation of data (where an individual was able to learn about protected information by combining several other less protected data elements). The threats and vulnerabilities that could lead to disclosure will be examined in section 1.6 in this chapter.

Need to know means that individuals are given access only to the information that is needed in order for them to perform their job functions. For example, most users do not need to see credit card numbers; therefore, the numbers are masked or hidden except for the last few digits. This protects the cardholder data from being compromised by a person that did not need to see the entire credit card number.

Least privilege is the restriction of data access of an individual or process to only the minimum level of access needed to perform their job functions. If a person has need to know access, then the next step is to determine what they can do with the data after they access them. For example, can they delete the data, or create or enter new data? Do they have read-only access or can they make changes or modifications?

It is important is to determine what level of impact would be realized if there was a breach of confidentiality. This determines the level of controls needed to protect the information from improper disclosure.

1.5.2 Integrity

Integrity is defined as the guarding against improper information modification, exclusion or destruction and includes ensuring information nonrepudiation and authenticity.

Maintaining integrity requires the protection of information from improper modification by internal authorized users, unauthorized users, or other processes or activities operating on the system. This requires that information is accurate to the extent required. It also protects the authenticity of information by identifying the sources of the data and ensuring that it has not been tampered with or destroyed. An example would be medical data. If a dosage of medication was incorrect, given to the wrong person or issued by an unauthorized person, the impact could be catastrophic. Integrity would be measured using the same levels as used when measuring the impact due to a breach of confidentiality.

1.5.3 Authenticity

Authenticity is often associated with integrity in relation to authenticating a person or process that is a participant in a transaction. This is often called nonrepudiation, and it ensures that a link can be made between an action and the source of the action. In public key infrastructure (PKI), for example, the ability to link an entity with their public key through the use of an X.509 certificate is essential to ensuring the authenticity of the transaction or the authenticity of the web site that is being accessed. A digital signature also proves both the authenticity of the document itself and the sender of the signed document.

1.5.4 Availability

Availability refers to providing timely and reliable access to information. Many business processes, such as industrial control systems (e.g., supervisory control and data acquisition [SCADA]), require data to be available in a timely, accurate manner so that controls can be monitored to protect safety and system operations. Availability is an excellent way to measure the importance or criticality of information and to document the importance of ensuring that data are accessible when required.

Availability can be measured using gap analysis. For example, when measuring network availability, it is possible to compare current levels of availability with required levels of availability. The gap between current and required mandates the projects, programs and initiatives required to close that gap.

When asking the business what levels of availability they require, the answer is often that the system and data are essential and require 100 percent availability. In most cases, this is not accurate, and this requires the risk practitioner to work with the business to determine what the true level of availability requirements are and what the organization is willing to pay to have higher availability. This will indicate the service level agreement (SLA) or key performance indicator (KPI) that the system is expected to meet.

Determination of the required level of availability is often found in a BIA. This is one source of information that can be referenced by the risk practitioner. After the required level of availability has been determined, the risk practitioner can recommend controls to obtain the required level of availability. In some cases, the system may be functioning at required levels or there may be a gap between the level of availability being provided and the level that is required or expected by the business. This gap indicates the effort needed to improve and update system availability to required levels. Addressing the gap may require a substantial effort and cost that cannot be addressed all at once. Therefore, decisions must be made by management on the priorities and steps necessary to resolve the outstanding issues. This may require several projects and initiatives to build the required levels of availability into the system. The completed business case should be presented to the risk owner so they can prioritize and determine the steps necessary to resolve the outstanding issues.

1.5.5 Segregation of Duties

Segregation of duties (SoD) is the principle of ensuring that no one person controls an entire transaction or operation that could result in fraudulent acts or errors. This is often enforced in financial transactions or cases where controls are needed to protect a system or operation. For example, SoD occurs when more than one person is required to open a bank vault or to approve a large payment.

When SoD is in place, incorrect or fraudulent data are more likely to be caught and the violation addressed. This can be implemented by requiring one person to review and approve the work of another person or by requiring two people to participate in a task simultaneously (often called dual control). SoD can be circumvented through collusion when the two people agree to bypass the SoD control.

The enforcement of SoD is often managed through the use of mutual exclusivity. Mutual exclusivity means that a person cannot execute both parts of the same transaction. They may be allowed to both input and approve data, but they would not be allowed to approve a transaction they input.

1.5.6 Job Rotation

If only one person in an organization has the required knowledge or ability to perform a critical job function or operate a system, this poses a risk to the organization in the event that the person is not available. Job rotation is the process of cross-training and developing personnel with various skills that can step in where needed. A risk associated with job rotation is that the training may make these employees more attractive to other employers, and the rotation of personnel may result in decreased efficiency at times of transition. Also, if employees know that they are about to move out of a job, they may not be diligent in doing their jobs properly.

1.5.7 Mandatory Vacations

Mandatory vacations are used in some financial organizations to detect fraud and are often required by law. A person who is in a position of trust in a bank or financial institution may have begun to issue fraudulent loans or mismanage the assets of the organization. Requiring that a person to take a mandatory vacation and having another person temporarily assume the responsibilities may allow for the detection of fraud or mismanagement.

1.5.8 Secure State

Secure state is the principle that a transaction or process should be in and maintain a secure condition at all times as it goes through its various activities. The consistent protection of a process is important to ensure that data are adequately protected at all times and that there is no time during a process that data or a system are vulnerable due to a lack of adequate security.

1.5.9 The IAAA Model of Access Control

One of the most critical areas of risk associated with information systems is the challenge of managing access control. Risk is often caused through misuse of access, especially in cases where a person has a level of access that is not appropriate for their current job responsibilities. Access control is usually addressed through the concepts of identification, authentication, authorization and accountability (IAAA).

Identification

Every person or process that uses a system should be uniquely identified. This allows tracking and logging of the activity by the user and the possibility to investigate a problem if it were to arise. Identification is often provided through a user ID, a customer account number, an employee identification number or other unique element. Sharing of user IDs should be prohibited, and the process of issuing an employee ID should be secure and require proper authorization.

Authentication

Authentication is the process of validating an identity. After a person or process has claimed or stated his/her identity, the process of authentication verifies that the person is who they say they are. Authentication ensures that one person cannot spoof an identity or masquerade as or impersonate another user. This can also prevent the sharing of user IDs. Authentication is usually done using three methods:
• Knowledge
• Ownership (possession)
• Characteristic (biometrics)

Authentication by knowledge requires users to know a password, code phrase or other secret value to validate their identity. The risk is that knowledge is often subject to replay attacks because the item used (password) can be used for an extended period of time. This allows a person learning the password of another person to log in as that user. This is why passwords should be changed on a regular basis.

Authentication by ownership requires the use of a smart card, token, ID badge or other similar item. This requires a person to validate their identity by having an access card or providing a one-time password that is generated by the token. The problem with ownership as a means of authentication is the cost of installing the system, issuing the cards, and operating and maintaining the system. In the event that a person loses his/her card, it may be used by an imposter if the card has not been reported as lost or stolen.

Biometrics uses either physiological (e.g., fingerprints, iris scan, palm scan, etc.) or behavioral (e.g., voice print, signature dynamics) elements to authenticate a person. Biometrics is expensive, and some users find it to be intrusive and may be resistant to it.

Node authentication differs from other types of authentication because it authenticates a device or location, not an individual or process. Node authentication is used when limiting access to certain devices through the identification of Internet protocol (IP) addresses, media access control (MAC) addresses or device serial numbers that are attempting to log on.

Strong Authentication

No single factor of authentication is strong enough on its own. All three forms can be bypassed fairly easily by a determined hacker. Therefore, it is recommended to use more than one type of authentication during the login process. This could include using a password with a smart card, or a token and biometrics as a multifactor authentication process. This is known as strong authentication.

Authorization

After a person or process has successfully logged on to a system, the system must provide that person or process the appropriate levels of authorization. Authorization refers to the privileges or permissions the person will have, including read-only, write-only, read/write, create, update, delete, full control, etc. This is where the concept of least privilege applies and ensures that employees only have the level of access they require to perform their job functions.

Temporal Isolation

A person's authorization is usually only granted for the period of time that the permissions are required. Most users can only access a building, network or application during normal working hours. Outside of those hours, access is not allowed. This prevents a user or other person from using the account at times when it should not be required.

Accountability/Auditing

The final part of the IAAA model is accountability or auditing. This action logs or records all activity on a system and indicates the user ID responsible for the activity. This underlines the need for using unique user IDs and ensuring that the logs are protected from modification or alteration. Logs are often needed for investigation and analysis and must be preserved to prevent a person from disabling the logging function or altering log data. Because logs may contain sensitive data, access to the logs should be strictly controlled.

1.5.10 Identity Management

Identity management is the process of managing the identities of the entities (users, processes, etc.) that require access to information or information systems. This is currently one of the toughest challenges for system administrators. Most people who have access to systems are outsiders—customers or guests—and they must be limited to only gaining access to appropriate information. There must also be a mechanism to reset passwords without creating a large amount of manual work.

New challenges occur when managing the access of internal employees, contractors and temporary employees. These users have a higher level of access than customers and guests, and their access must be removed when it is no longer required. Their access should also be managed to prevent the accumulation of access over time, so that users do not have access that far exceeds their current requirements. This often happens when a person has worked for an organization for an extended time period and has moved from one job to another—each job requiring different access permissions.

Access permissions that are no longer in regular use but are left on a system are a risk because they may be used by another person or an attacker. This use will not be readily noticed because no one else is using that account.

1.6 THREATS AND VULNERABILITIES RELATED TO ASSETS (PEOPLE, PROCESSES AND TECHNOLOGY)

Identification of risk is related to the identification of assets, threats to those assets and the vulnerabilities that those assets contain. Key terms for the identification of threats and vulnerabilities are listed in **exhibit 1.14**.

Exhibit 1.14: Risk Identification Terminology	
Term	**Definition**
Asset	Something of either tangible or intangible value that is worth protecting, including people, information, infrastructure, finances and reputation
Asset value	The value of an asset is subject many factors including the value to both the business and to competitors. An asset may be valued according to what another person would pay for it, or by its measure of value to the company. Asset value is usually done using a quantitative (monetary) value
Impact	Magnitude of loss resulting from a threat exploiting a vulnerability
Impact analysis	A study to prioritize the criticality of information resources for the enterprise based on costs (or consequences) of adverse events. In an impact analysis, threats to assets are identified and potential business losses determined for different time periods. This assessment is used to justify the extent of safeguards that are required and recovery time frames. This analysis is the basis for establishing the recovery strategy
Impact assessment	A review of the possible consequences of a risk
Likelihood	The probability of something happening
Threat	Anything (e.g., object, substance, human) that is capable of acting against an asset in a manner that can result in harm
Threat agent	Methods and things used to exploit a vulnerability
Threat analysis	An evaluation of the type, scope and nature of events or actions that can result in adverse consequences; identification of the threats that exist against enterprise assets
Threat vector	The path or route used by the adversary to gain access to the target
Vulnerability	A weakness in the design, implementation, operation or internal control of a process that could expose the system to adverse threats from threat events
Vulnerability analysis	A process of identifying and classifying vulnerabilities
Vulnerability scanning	An automated process to proactively identify security weaknesses in a network or individual system

1.6.1 Risk Factors

Risk is a combination of several factors that interact to cause damage to the assets of the organization. The relationships between risk factors are described in **exhibit 1.15**.

Through the identification of risk, the risk practitioner will be able to assess the likelihood and impact of various risk events that could impact an asset.

Threat agents (e.g., hackers) use threats (e.g., viruses) to attack an asset via a vulnerability. This indicates the risk that the asset faces. If the attacker does not have the capability (a tool [threat]), then they are not able to pose a significant risk. Likewise, if the asset is properly protected and is not vulnerable to the threat being issued, no real risk exists. For example, if a virus has been written to exploit a vulnerability on a system, the virus can only have an impact if the system has not been patched. After the patch has been applied, the virus poses no real threat to the system. The process of recognizing what can go wrong and why it could happen is called threat and vulnerability analysis.

Knowledge of threats and the motivations, strategy and techniques of those who perpetrate threats is key, so that a threat may be managed before it becomes reality. The better understanding the risk practitioner has of the mind of the attacker or the source of the threat, the more effective the risk management activities will be in controlling the threat.

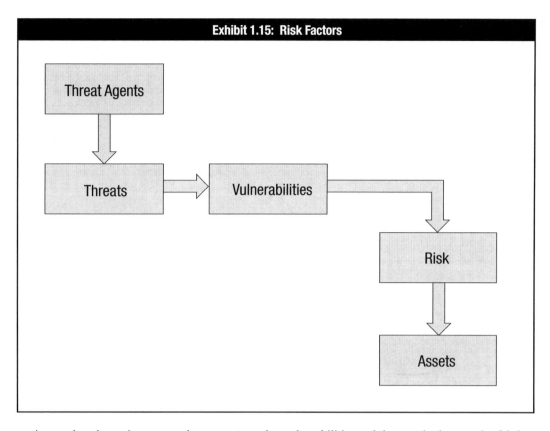

An enterprise needs to know its own weaknesses, strengths, vulnerabilities and the gaps in the security fabric. Security must be woven into each business process, each IT system and all operational procedures. Any gaps in the security fabric can easily be exploited.

The threat and vulnerability landscape is always changing. Personnel move, equipment deteriorates, controls are weakened, new threats emerge and the awareness of security may become dulled. The risk practitioner must proactively seek out threats, measure vulnerabilities and assess controls through regular assessments, testing, observation and analysis. Performing a vulnerability assessment on an information system on a scheduled basis or when a change occurs can be instrumental in the organization finding a vulnerability before an adversary does and preventing an attack that may otherwise occur.

Risk is often considered with a narrow framework that focuses on equipment or natural disasters, but disregards other risk factors of people and processes. When evaluating risk, it is important to consider all types of risk, including technology-related risk. This type of risk can include a threat actor intercepting, interrupting, modifying or fabricating data on information assets, as well as the risk to the people and procedures that are used in acquiring, implementing, maintaining and disposing of that technology.

Risk is often influenced more by lack of training than by lack of equipment. In many cases, the risk is related to the way the equipment is operated more than whether the right tools are available.

Risk evaluation is the measurement of risk, and the entire risk environment must be evaluated. The risk environment includes:
• The context, criticality and sensitivity of the system or process being reviewed
• The dependencies and requirements of the system or process being reviewed
• The operational procedures, configuration and management of the system or technology
• The training of the users and administrators
• The effectiveness of the controls and monitoring of the system or business process
• The manner in which data and system components are decommissioned

1.6.2 Elements of Risk

Risk identification requires the documentation and analysis of the elements that comprise risk. Risk only occurs if the adversary has intent (motivation) and capability. Risk is based on the value of the asset being threatened, the presence of threats and vulnerabilities, and the likelihood that a threat may exploit a vulnerability either accidentally or intentionally and cause harm. Each of the elements of risk needs to be considered both individually and in aggregate.

1.6.3 Assets

An asset is something of either tangible or intangible value that is worth protecting. Assets include:
• Information
• Reputation
• Brand
• IP
• Facilities
• Equipment
• Cash and investments
• Customer lists
• Research
• People
• Service/business process

To implement a justifiable risk treatment strategy, it is necessary to start by identifying the organization's assets and determining the value of those assets. Not all assets are equally important; many are absolutely critical to business operations, and others are just a convenience. Obviously, priority should be given to protecting the most important assets first and then addressing the risk requirements of less significant assets as time and budget allow. In the end, an organization must know the value of their assets to avoid paying more in protection than the net worth of the asset.

The calculation of asset value is not as straightforward as it may initially appear. Many organizations use a quantitative approach that puts a monetary value on an asset. This can be difficult because the value of an asset can also be dependent on intangibles such as confidence, morale or market perception. For example, if a product, company or region is associated with poor quality, environmental negligence or fraudulent activity, a high-quality product may be perceived as substandard due to this market perception. This negative perception may persist for many years before the company is able to recover.

When calculating asset value, one technique is to base it on the impact of a loss of CIA. This approach attempts to relate the impact to easily understood terminology. In order for this to work, the values for each level must be clearly stated and used in the same way by all departments.

Contributing factors to calculating asset value include:
• Financial penalties for legal noncompliance
• Impact on business processes
• Damage to reputation
• Additional costs for repair/replacement
• Effect on third parties and business partners
• Injury to staff or other personnel
• Violations of privacy
• Breach of contracts
• Loss of competitive advantage
• Legal costs

1.6.4 Threats

Despite best efforts, threats will always exist. Threats are often beyond the direct control of the risk practitioner or the owner of the asset. Threats can be external or internal and intentional or unintentional. They may be caused by natural events or political, economic or competitive factors. Risk should be tailored for each organization, and not all threats will be a factor for every organization. For example, an organization that operates in a region with a seismic rating of zero does not have to document their exposure to volcanoes or earthquakes. They may note that those threats were intentionally bypassed just to assure an audit that the threat was not missed or forgotten. The risk practitioner may create a threat assessment report to document the results of the threat analysis.

It is important to identify the various types of threats and ensure that the enterprise understands the vectors or angles of attack that a threat agent may use to compromise the systems.

Generally, threats are divided into several categories including:[9]
- Physical
- Natural events
- Loss of essential services
- Disturbance due to radiation
- Compromise of information
- Technical failures
- Unauthorized actions
- Compromise of functions

Several excellent lists of threats are available that can be used by a risk practitioner to ensure that all threats have been considered. These include the *NIST Special Publication 800-30 Revision 1: Guide to Conducting Risk Assessments*, available at *csrc.nist.gov*, and ISO/IEC 27005:2008 Appendix C available at *iso.org*.

The risk practitioner must document all the threats that may apply to the systems and the business processes that are under review. This requires using the resources noted previously, but should also include examining the cause of past failures, audit reports, media reports, information from national computer emergency response teams (CERTs), data from security vendors and communication with internal groups. An attacker is often imaginative, creative and determined and will explore new methods and avenues of attack. The risk practitioner must be equally determined and creative and seek to discover as many threats as possible. If a threat or threat vector (i.e., the way the threat attacks) has not been identified, then the organization may be unprepared and vulnerable to that threat.

Threats may be the result of accidental actions, intentional/deliberate actions or natural events. Threats may originate from either internal or external sources, and actual attacks may leverage a combination of both internal and external sources.

Sources for information regarding threats are listed in **exhibit 1.16**.

Exhibit 1.16: Sources for Threat Information	
Service providers	Insurance companies
Threat monitoring agencies	Product vendors
Security companies	Government publications
Audits	Assessments
Management	Users
Business continuity	Human resources
Finance	Media

Internal Threats

Personnel are often considered an organization's best asset. However, in many organizations, employees may be unhappy because they are inadequately trained, treated poorly or not given enough time to do their jobs properly. This leads to a higher risk of errors, negligence, theft, system failure and costs. Key personnel may be drawn to another company or may get frustrated and leave. This may leave a serious gap in the knowledge and skills needed to operate systems effectively.

Employees are the cause of a significant number of attacks. A disgruntled employee may intentionally cause harm or may leak important IP. An employee may be convinced, bribed or threatened to disclose secret data or information for ideological or economic reasons. The solution to the employee problem is difficult to find. Many employees have a level of access to systems and data that far exceeds their actual job requirements, and this vulnerability can be exploited in an attack.

The trusted or malicious insider threat to an organization is a current or former employee, contractor, or other business partner who has or had authorized access to an organization's network, system or data and intentionally intercepted (exfiltrated), interrupted, modified or fabricated data on the organization's information systems.

The first step in addressing personnel threats is to start with the hiring process and to review the qualifications and attitude of prospective employees. Employment candidates may have submitted incorrect information on job applications and claimed education, certification or experience that they did not actually possess. At the time of hiring, the employee should be required to sign a nondisclosure agreement and be advised of the ethics and policies of the organization. In some cases, a review of references and background checks can be a wise decision.

During employment, employees should be reminded of organizational policies and their responsibilities through awareness sessions and regular management reviews. One of the best employee-based controls is to interact with employees to understand any frustrations, complaints or issues that they may be facing and to seek to resolve those issues. During times of strike, layoffs, mergers, relocation and reorganization, an employee is more likely to be a risk. An employee who has been recently demoted or bypassed for a promotion is also a risk.

At the end of employment, an employee should return all assets including identification badges, equipment (e.g., laptops, mobile phones, access cards, etc.) and uniforms so that he/she cannot use those to gain unauthorized access in the future. All systems, network and facility access should be removed immediately on the employee's departure.

External Threats

Threats to information systems can include (but are not limited to):
• Espionage
• Theft
• Sabotage
• Terrorism
• Criminal acts
• Software errors
• Hardware flaws
• Mechanical failures
• Lost assets
• Data corruption
• Facility flaws (freezing/pipe burst)
• Fire
• Supply chain interruption
• Industrial accidents
• Disease (epidemic)
• Seismic activity
• Flooding
• Power surge/utility failure
• Severe storms

Natural events such as flood, storm, earthquake or tornado are unpredictable and may be extremely damaging. The use of governmental data and weather monitoring services may identify the threats associated with natural events and allow the risk practitioner to take necessary steps to be prepared.

An external personnel threat includes a hacker, a thief or an advanced persistent threat (APT) that is skilled and determined to break into systems for military or economic purposes. APTs are advanced, highly skilled attackers that are determined (persistent) in their attempts to exploit systems and networks. The increased skills available to the hacking community and the effectiveness of the tools they possess make the risk of compromise much more significant. Many APTs are sponsored by governments, organized crime or competitors.

It is reported that most breaches are the result of targets of opportunity, not determined attacks. As seen in the annual reports from Verizon[10] and other organizations, the reason most organizations are breached is because they were discovered to be an easy target and the hacker merely took advantage of their vulnerabilities.

Emerging Threats

Indications of emerging threats may also include unusual activity on a system, repeated alarms, slow system or network performance, or new or excessive activity in logs. In many cases, an organization that has experienced an unexpected breach had evidence of the emergence of the threat in their logs prior to the breach, but the evidence was either not noticed or acted on. The lack of monitoring a vulnerability, when combined with a threat, is often the combination of factors that lead to a breach.

New technologies are often a new threat source because most technologies are built with an emphasis on function and purpose without due consideration for the security implications associated with the new technology. The risk practitioner must be alert to the emergence of new technologies and prepare for the introduction of those technologies into the organization. It would be naive to assume that employees and management are not bringing those technologies into the organization as soon as they are available.

The risk practitioner should seek to ensure that all threats have been considered and documented. However, it is also good to remember that risk should be tailored for each organization and not all threats will be a factor for every organization. This includes modifying the threat and vulnerability assessment according to the needs of the organization, the operational environment (e.g., operating in a secure facility versus operating in an insecure facility) and the value of the assets. Some systems may not be networked, or they may only be connected to an isolated system, so the risk is different for Internet-based systems. The risk practitioner may note that those threats were intentionally bypassed to assure an audit that the threat was not missed or forgotten. The risk practitioner may create a threat assessment report to document the results of the threat analysis.

1.6.5 Vulnerabilities

An enterprise needs to be familiar with vulnerabilities, such as the weaknesses, gaps or potential holes in their security fabric, that could allow an attack to succeed.

NIST Special Publication 800-30 Revision 1: Guide to Conducting Risk Assessments provides a list of vulnerabilities to consider. It also uses the term "predisposing conditions." This highlights the concept that many vulnerabilities are conditions that exist in systems and must be identified so that they can be addressed. The purpose of vulnerability identification is to find the problems before they are found by an adversary and exploited. This is why an organization should conduct regular vulnerability assessments and penetration tests to identify, validate and classify any vulnerabilities.

NOTE: Other sources of vulnerabilities include:
- National Vulnerability Database at *nvd.nist.gov*
- Common Weakness Enumeration at *nvd.nist.gov/cwe.cfm*
- Harmonized Threat and Risk Assessment at *www.cse-cst.gc.ca*
- Contingency Planning and Management at *contingencyplanning.com*
- Open Web Application Security Project (OWASP) at *www.owasp.org*

Network Vulnerabilities

Network vulnerabilities are often related to misconfiguration of equipment, poor architecture or traffic interception. Misconfiguration is a common problem with network equipment that is not properly installed, operated or maintained. Network equipment should be hardened to disable any unneeded services, ports or protocols. Any open services are a potential attack vector that can be exploited by an attacker. Many compromises of organizations' networks are facilitated by an unneeded open port that should have been disabled. One possible reason for this is that network engineers may not have received sufficient training to know how to properly manage network devices.

Physical Access

The *Verizon 2013 Data Breach Investigations Report*[11] indicated that physical security was a factor in 35 percent of all data breaches. A vulnerability in physical security has the potential to bypass nearly every other type of control. That is why physical security is so important. The ability to access server rooms, network cabling, information systems equipment and buildings can allow an attacker to circumvent passwords, steal equipment, intercept data communications and take ownership of nearly any system or device. Testing for physical security vulnerabilities includes testing locks, security guards, fire suppression systems, heating ventilation and air conditioning controls, and lighting, cameras and motion sensors.

Applications and Web-facing Services

Applications, especially web applications, are one of the most common entry points currently used by attackers. Many applications are written with a focus on supporting a business function without due regard to the security requirements of the applications. In many cases, the application may be vulnerable to buffer overflows, logic flaws, injection attacks, bugs, incorrect control over user access and many other common vulnerabilities. The risk practitioner can use tools from the Open Web Application Security Project (OWASP)[12] to test web-facing applications for well-known and documented vulnerabilities. These vulnerabilities can lead to data compromise or system failure.

Another reason for application-related breaches is poor architecture. The application may process payment card data or other sensitive information, but store that data in an insecure manner or an insecure location, such as in a demilitarized zone (DMZ).

Utilities

Information systems rely on controlled environmental conditions including clean and steady power and controls over humidity and temperature. The risk manager must be prepared for power failure or other environmental conditions that may lead to system failure. Having an uninterruptible power supply (UPS), backup generators and surge protectors can protect equipment from damage or failure. A UPS must be validated to ensure it has adequate power to run critical systems, at least until a backup generator can come online.

Supply Chain

Many organizations rely on products, raw materials and supplies from around the world, and any interruption in the supply chain could affect their ability to function. For example, a shortage in supply of fuel to an airport would mean that all operations are impacted and the business would not be able to meet its goals.

Processes

Maintaining operational integrity requires the use of procedures such as incident management, identity management, change control, patch management and project planning. These processes must be defined and deployed in a consistent manner across the organization. Without these controls, the organization may be at risk of inconsistent management and results, lack of governance and reporting, and failure to ensure compliance with regulations.

Equipment

As equipment ages, it becomes less efficient, less effective and unable to support modern business functions. Equipment is often sold with a mean time between failure (MTBF). This is an indication of the anticipated life span of the device and indicates when the device should be scheduled for removal/replacement.

Cloud Computing

The use of cloud-based services is very attractive to many organizations for several reasons, including flexibility, cost savings, availability of expert support and the ability to outsource a noncore function. There are several cloud-based service models, as seen in **exhibit 1.17**.

Exhibit 1.17: Cloud Deployment Models	
Private cloud	• Operated solely for an enterprise • May be managed by the enterprise or a third party • May exist on- or off-premise
Public cloud	• Made available to the general public or a large industry group • Owned by an organization selling cloud services
Community cloud	• Shared by several enterprises • Supports a specific community that has a shared mission or interest • May be managed by the enterprises or a third party • May reside on- or off-premise
Hybrid cloud	• A composition of two or more clouds (private, community or public) that remain unique entities, but are bound together by standardized or proprietary technology that enables data and application portability (e.g., cloud bursting for load balancing between clouds)
Source: ISACA, *IT Control Objectives for Cloud Computing: Controls and Assurance in the Cloud*, USA, 2011, figure 1.3	

The outsourcing of data processing does not remove the liability of the outsourcing organization to ensure that its data are properly protected and that the storage and transmission of data is compliant with laws on data transfer.

Big Data

Advances in the capability to perform analysis of data from various sources—both structured and unstructured data—allow enterprises to make better business decisions and increase competitive advantage. This change in analytics capabilities, termed "big data analytics," can introduce additional technical and operational risk. Enterprises should understand that risk can be incurred either through adoption or nonadoption of big data analytics; specifically, enterprises must weigh the technical and operational risk against the business risk that is associated with failure to adopt.

The risk of adopting or not adopting big data analytics is described by ISACA the following way:[13]

Technical and operational risk should consider that certain data elements may be governed by regulatory or contractual requirements and that data elements may need to be centralized in one place (or at least be accessible centrally) so that the data can be analyzed. In some cases, this centralization can compound technical risk.

For example:
• Amplified technical impact—If an unauthorized user were to gain access to centralized repositories, it puts the entirety of those data in jeopardy rather than a subset of the data.
• Privacy (data collection)—Analytics techniques can impact privacy; for example, individuals whose data are being analyzed may feel that revealed information about them is overly intrusive.
• Privacy (re-identification)—Likewise, when data are aggregated, semianonymous information or information that is not individually identifiable information might become non-anonymous or identifiable in the process.

Vulnerability Assessment

A vulnerability assessment is a careful examination of a target environment to discover any potential points of compromise or weakness. It may be a manual process or use automated tools. A manual test is better for some types of tests, such as physical security checks or source code validation. However, an automated tool has the ability to filter large amounts of data and can be used to examine logs, compile and analyze data from multiple sources (e.g., a security event and incident management [SEIM] tool), examine the functions of a program (e.g., black box and white box testing), and run test files or data against a tool such as a firewall or application.

Vulnerabilities include:
- Network vulnerabilities
- Poor physical access controls (e.g., buildings, offices)
- Insecure applications
- Poorly designed or implemented web-facing services
- Disruption to utilities (e.g., power, telecommunications)
- Unreliable supply chain
- Untrained personnel (human resources)
- Inefficient processes (e.g., change control, incident handling)
- Poorly maintained or old equipment

1.6.6 Penetration Testing

The results of vulnerability assessments should be combined into a report. The challenge is that a vulnerability assessment may contain noise or data that is not accurate. A vulnerability that is not truly a problem may appear on a vulnerability assessment report. This is a why an organization will often follow a vulnerability assessment with a penetration test.

Penetration testing is a targeted test against a potential vulnerability or against an attack vector commonly used by an attacker that simulates the activities and approach used by an attacker. Penetration testing uses the same sort of tools as those used by a real adversary. A penetration test validates whether an identified vulnerability is a true weakness or whether it is a false positive (an alert for a condition that does not exist).

An internal or external team can conduct a penetration test. Types of penetration tests range from a full-knowledge test, where the testing team is familiar with the entire infrastructure that is being tested, to a zero-knowledge test, where the testing team is in the position of an external hacker who has no knowledge of the infrastructure being attacked. An expert penetration testing team will be creative and attempt several types of tests to ensure that as many attack vectors as possible have been tested.

At the end of the test, the team should submit a report that lists the types of tests conducted, the results of the tests and recommendations of possible ways to mitigate any identified vulnerabilities.

> **Additional information on penetration testing can be found in chapter 2, section 2.3.12 and chapter 4, section 4.6.2.**

1.6.7 Likelihood/Probability

The results of vulnerability assessments are a good indication of the likelihood that an organization will face a successful attack. The presence of a vulnerability will not guarantee an attack, but it is much more likely and probable that a vulnerability will be discovered and exploited eventually by an adversary. The likelihood of an attack is often a component of external factors, such as the motivation of the attacker, as shown in **exhibit 1.18**.

Motivation

The likelihood of an attack is influenced by the attacker's motivation. A disgruntled, unhappy employee; an employee who is about to lose his/her job; a thief desperate for money; or a foreign government trying to protect its interests are examples of key motivators. The more motivated the attacker is, the more likely it is that he/she will attack, and the more persistent he/she may be.

Skills

Other factors that influence likelihood include the skills and tools available to the attacker. If an attacker does not have the tools necessary to launch the attack or the skill needed to defeat the protective controls of the target, the attack is unlikely to succeed.

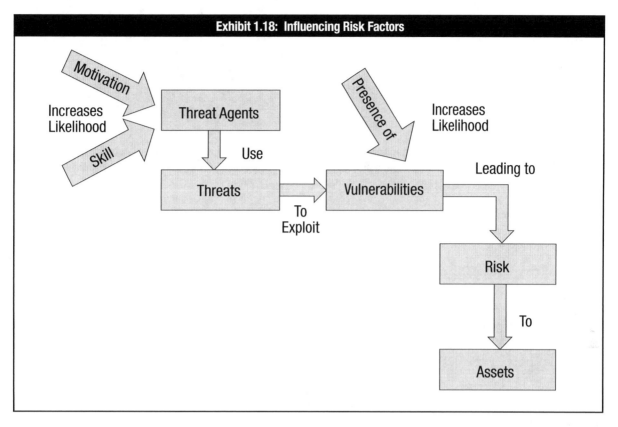

Exhibit 1.18: Influencing Risk Factors

Presence of a Vulnerability

A secure system is far less vulnerable to an attack than an insecure system, and the presence of vulnerabilities, especially of known vulnerabilities, will make the likelihood of a successful attack much greater. The presence of a vulnerability, therefore, is a factor in the calculation of likelihood.

1.7 IT RISK RELATED TO ORGANIZATIONAL ASSETS AND BUSINESS PROCESSES

For most organizations, the most valuable asset is either people or information. The greatest risk, therefore, is the risk of loss or corruption of that information, or the loss of key personnel. The risk practitioner should identify the areas of risk associated with these areas.

1.7.1 People

Many organizations are vulnerable to the loss of a key employee who may be the only person with knowledge in a certain area or specific expertise. The loss of an employee through retirement, illness or recruitment by another organization may leave the organization in a very precarious and vulnerable position. The failure of management to identify key employees and ensure that they are supported through cross-training and incentive programs is far too common. Even when a departure is expected, such as an employee's retirement, organizations can still find themselves unprepared.

1.7.2 Technology

Technology is changing rapidly, and new technologies are always being developed. The risk practitioner should be aware of new technologies and the risk they pose; however, there are other concerns that may be overlooked related to the technology already in use.

Outdated technology is often an overlooked risk in an organization. Equipment that is past its MTBF is often vulnerable. The lack of patching and updating of systems and applications leaves them vulnerable to malware or misuse. Older systems may require expertise that is not readily available to maintain. Older systems may not be documented may be difficult to obtain replacement parts for or may be reliant on an individual to maintain.

At end of life, system hardware may contain sensitive data that must be securely deleted. Often this will require overwriting, degaussing or physical destruction of the equipment. The method of secure disposal will be based on the risk associated with the data on the device. There is also the risk that data may be needed later on and the organization may need to retain a copy of the software to read the data. If the data are encrypted, the keys must also be stored. Failure to remove the system from backups or business continuity or disaster recovery plans may also affect the integrity of those operations.

1.7.3 Data

Data are a valuable asset of the organization. Customer lists, financial data, marketing plans, human resources data and research are some of the data-related assets that must be protected. Sensitive data must be protected from disclosure or modification; critical data must be protected from destruction or loss. The systems that host, process or transmit the data must ensure that data are protected at all times, in all forms (paper, magnetic storage, optical storage, reports, etc.) and in all locations (storage, networks, filing cabinets, archives, etc.).

1.7.4 Trademarks and Intellectual Property

A special form of information is IP. This includes trademarks, copyrights, patents, brands and other items associated with the reputation and goodwill of the organization. Research that leads to a new product may represent the future earnings potential of the organization, and failure to protect it from disclosure may result in the loss of competitive advantage or future earnings. All employees and business partners should be bound by nondisclosure agreements (NDAs) and reminded of their responsibility to protect the IP of the organization and handle it properly. This may include strict access controls, shredding of documents, caution when discussing information in a public location and encryption of data on portable media. The controls will be examined in more detail in chapter 3, Risk Response and Mitigation.

The risk associated with IP is that failure to protect the IP from improper use, disclosure or duplication may result in the loss of the IP protection for that product. This would allow a competitor to use the IP without compensating the original owner. Key terms related to IP are described in **exhibit 1.19**.

Exhibit 1.19: Intellectual Property Terms	
Term	**Definition**
Trademark	A sound, color, logo, saying or other distinctive symbol that is closely associated with a certain product or company
Copyright	Protection of writings, recordings or other ways of expressing an idea
Patent	Protection of research and ideas that led to the development of a new, unique and useful product to prevent the unauthorized duplication of the patented item

1.7.5 Risk Related to Business Processes

A risk related to business processes is inefficient or outdated processes that have not been changed or updated. An organization may continue to do things "the way they have always done" them rather than move to an updated process. This may result in the organization becoming noncompetitive in the future. A business process must be flexible enough to adapt to changes in the market or technology. An organization must be able to understand the effect of a change in market conditions, customer expectations or regulations.

1.8 IT RISK SCENARIOS

To properly assess risk in a qualitative manner, it is necessary to develop risk scenarios that will be used in the IT risk assessment. Each scenario should be based on an identified risk, and each risk should be identified in one or more a scenarios. Each scenario is used to document the level of risk associated with the scenario in relation to the business objectives or operations that would be impacted by the risk event.

COBIT 5 for Risk describes a risk scenario as:

> *A risk scenario is a description of a possible event that, when occurring, will have an uncertain impact on the achievement of the enterprise's objectives. The impact can be positive or negative.*

> *The core risk management process requires risk needs to be identified, analysed and acted on. Well-developed risk scenarios support these activities and make them realistic and relevant to the enterprise.*[14]

1.8.1 Risk Scenario Development Tools and Techniques

The development of the risk scenarios is an art. It requires creativity, thought, consultation and questioning. If an incident has occurred previously, it does not require much thought or preparation to document the risk event, unless the event was not examined in detail. In many cases, the risk practitioner finds that the first concern of the management team was to treat the event as a rare event rather than thoroughly delve into the specifics of the event. This will result in a risk scenario that may easily happen again and that could have been avoided through a better approach to investigating and resolving the issues associated with the event.

The development of risk scenarios is based on describing a potential risk event and documenting the factors and areas that may be affected by the risk event. Risk events may include system failure, loss of key personnel, theft, network outages, power failures, natural disasters or any other situation that could affect business operations and mission. Each risk scenario should be related to a business objective or impact.

The key to developing effective scenarios is to focus on real and relevant potential risk events. Examples of this would be to develop a risk scenario based on a radical change in the market for an organization's products, a change in government or leadership, or a supply chain failure.

There are several different approaches for the development of risk scenarios. As seen in **exhibit 1.20**, a risk scenario can be developed from a top down or from a bottom up perspective.

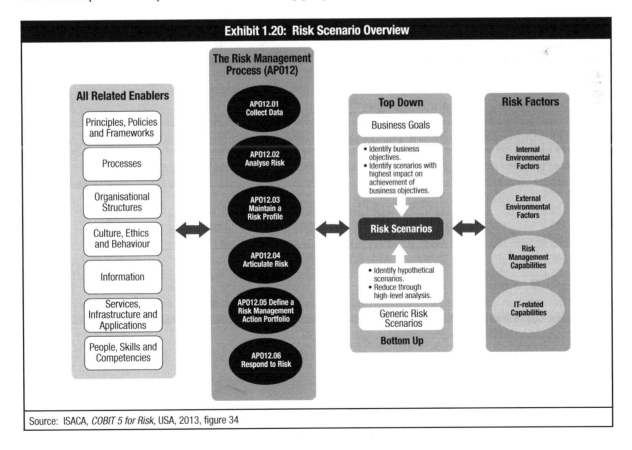

Exhibit 1.20: Risk Scenario Overview

Source: ISACA, *COBIT 5 for Risk*, USA, 2013, figure 34

Exhibit 1.21 shows an example of the many inputs that are required to develop risk scenarios.

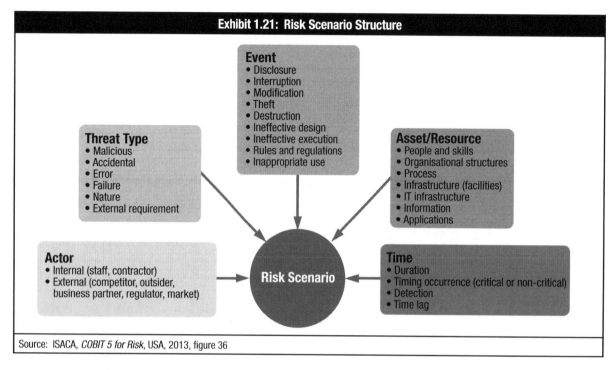

Exhibit 1.21: Risk Scenario Structure

Event
- Disclosure
- Interruption
- Modification
- Theft
- Destruction
- Ineffective design
- Ineffective execution
- Rules and regulations
- Inappropriate use

Threat Type
- Malicious
- Accidental
- Error
- Failure
- Nature
- External requirement

Asset/Resource
- People and skills
- Organisational structures
- Process
- Infrastructure (facilities)
- IT infrastructure
- Information
- Applications

Actor
- Internal (staff, contractor)
- External (competitor, outsider, business partner, regulator, market)

Risk Scenario

Time
- Duration
- Timing occurrence (critical or non-critical)
- Detection
- Time lag

Source: ISACA, *COBIT 5 for Risk*, USA, 2013, figure 36

Top Down Approach

A top down approach to scenario development is based on understanding business goals and how a risk event could affect the achievement of those goals. The risk practitioner looks for the outcome of events that may hamper business goals. Various scenarios are developed that allow the organization to examine the relationship between the risk event and the business goals, so that the impact of the risk event can be measured. By directly relating a risk scenario to the business, management can be educated and involved in how to understand and measure risk.

The top down approach is suited to general risk management of the company, because it looks at both IT- and non-IT-related events. A benefit of this approach is that because it is more general, it is easier to achieve management buy-in even if management usually is not interested in IT.

Bottom Up Approach

The bottom up approach to developing risk scenarios is based on the describing risk events that are specific to individual enterprise situations. An example of this would be developing scenarios based on a risk event in one department, on a failure of one IT system or a failure of a business process.

1.8.2 Developing IT Risk Scenarios

A risk scenario is a description of an IT-related risk event that can lead to a business impact. The risk scenario contains the following components:
- Actor: The internal or external party or entity that generates the threat
- Threat type: The nature of the threat event (malicious or accidental; a natural event; an equipment or process failure)
- Event: The security incident, such as the disclosure of information, the interruption of a system or project, including:
  - Theft
  - Improper modification of data or a process
  - Inappropriate use of resources
  - Changes to regulations
  - Lack of change management

- Asset: The entity affected by the risk event, including:
  - People
  - Organizational structure
  - IT processes
  - Physical infrastructure
  - IT infrastructure
  - Information, applications
- Time: If relevant to the scenario, including:
  - Duration (extended outage)
  - Timing (at a critical moment)
  - Detection (immediate detection or not)
  - Time lag between the event and consequence (immediate impact of an network failure versus long term problems from poor infrastructure)

Some of the key issues related to the development of risk scenarios are addressed in **exhibit 1.22**.

Exhibit 1.22: Risk Scenario Technique Main Focus Areas	
Focus/Issue	**Summary Guidance**
Maintain currency of risk scenarios and risk factors.	Risk factors and the enterprise change over time; hence, scenarios will change over time, over the course of a project or over the evolution of technology. For example, it is essential that the risk function develop a review schedule and the CIO works with the business lines to review and update scenarios for relevance and importance. Frequency of this exercise depends on the overall risk profile of the enterprise and should be done at least on an annual basis, or when important changes occur.
Use generic risk scenarios as a starting point and build more detail where and when required.	One technique of keeping the number of scenarios manageable is to propagate a standard set of generic scenarios through the enterprise and develop more detailed and relevant scenarios when required and warranted by the risk profile only at lower (entity) levels. The assumptions made when grouping or generalising should be well understood by all and adequately documented because they may hide certain scenarios or be confusing when looking at risk response. For example, if 'insider threat' is not well defined within a scenario, it may not be clear whether this threat includes privileged and non-privileged insiders. The differences between these aspects of a scenario can be critical when one is trying to understand the frequency and impact of events, as well as mitigation opportunities.
Number of scenarios should be representative and reflect business reality and complexity.	Risk management helps to deal with the enormous complexity of today's IT environments by prioritising potential action according to its value in reducing risk. Risk management is about reducing complexity, not generating it; hence, another plea for working with a manageable number of risk scenarios. However, the retained number of scenarios still needs to accurately reflect business reality and complexity.
Risk taxonomy should reflect business reality and complexity.	There should be a sufficient number of risk scenario scales reflecting the complexity of the enterprise and the extent of exposures to which the enterprise is subject. Potential scales might be a 'low, medium, high' ranking or a numeric scale that scores risk importance from 0 to 5. Scales should be aligned throughout the enterprise to ensure consistent scoring.
Use generic risk scenario structure to simplify risk reporting	Similarly, for risk reporting purposes, entities should not report on all specific and detailed scenarios, but could do so by using the generic risk structure. For example, an entity may have taken generic scenario 15 (project quality), translated it into five scenarios for its major projects, subsequently conducted a risk analysis for each of the scenarios, then aggregated or summarised the results and reported back using the generic scenario header 'project quality'.

Exhibit 1.22: Risk Scenario Technique Main Focus Areas *(cont.)*	
Focus/Issue	**Summary Guidance**
Ensure adequate people and skills requirements for developing relevant risk scenarios.	Developing a manageable and relevant set of risk scenarios requires: • Expertise and experience, to not overlook relevant scenarios and not be drawn into highly unrealistic or irrelevant scenarios. While the avoidance of scenarios that are unrealistic or irrelevant is important in properly utilising limited resources, some attention should be paid to situations that are highly infrequent and unpredictable, but which could have a cataclysmic impact on the enterprise. • A thorough understanding of the environment. This includes the IT environment (e.g., infrastructure, applications, dependencies between applications, infrastructure components), the overall business environment, and an understanding of how and which IT environments support the business environment to understand the business impact. • The intervention and common views of all parties involved—senior management, which has the decision power; business management, which has the best view on business impact; IT, which has the understanding of what can go wrong with IT; and risk management, which can moderate and structure the debate amongst the other parties. • The process of developing scenarios usually benefits from a brainstorming/workshop approach, where a high-level assessment is usually required to reduce the number of scenarios to a manageable, but relevant and representative, number.
Use the risk scenario building process to obtain buy-in.	Scenario analysis is not just an analytical exercise involving 'risk analysts'. A significant additional benefit of scenario analysis is achieving organisational buy-in from enterprise entities and business lines, risk management, IT, finance, compliance and other parties. Gaining this buy-in is the reason why scenario analysis should be a carefully facilitated process.
Involve first line of defence in the scenario building process.	In addition to co-ordinating with management, it is recommended that selected members of the staff who are familiar with the detailed operations be included in discussions, where appropriate. Staff whose daily work is in the detailed operations are often more familiar with vulnerabilities in technology and processes that can be exploited.
Do not focus only on rare and extreme scenarios.	When developing scenarios, one should not focus only on worst-case events because they rarely materialise, whereas less-severe incidents happen more often.
Deduce complex scenarios from simple scenarios by showing impact and dependencies.	Simple scenarios, once developed, should be further fine-tuned into more complex scenarios, showing cascading and/or coincidental impacts and reflecting dependencies. For example: • A scenario of having a major hardware failure can be combined with the scenario of failed DRP. • A scenario of major software failure can trigger database corruption and, in combination with poor data management backups, can lead to serious consequences, or at least consequences of a different magnitude than a software failure alone. • A scenario of a major external event can lead to a scenario of internal apathy.
Consider systemic and contagious risk.	Attention should be paid to systemic and/or contagious risk scenarios: • **Systemic**—Something happens with an important business partner, affecting a large group of enterprises within an area or industry. An example would be a nationwide air traffic control system that goes down for an extended period of time, e.g., six hours, affecting air traffic on a very large scale. • **Contagious**—Events that happen at several of the enterprise's business partners within a very short time frame. An example would be a clearinghouse that can be fully prepared for any sort of emergency by having very sophisticated disaster recovery measures in place, but when a catastrophe happens, finds that no transactions are sent by its providers and hence is temporarily out of business.
Use scenario building to increase awareness for risk detection.	Scenario development also helps to address the issue of detectability, moving away from a situation where an enterprise 'does not know what it does not know'. The collaborative approach for scenario development assists in identifying risk to which the enterprise, until then, would not have realised it was subject to (and hence would never have thought of putting in place any countermeasures). After the full set of risk items is identified during scenario generation, risk analysis assesses frequency and impact of the scenarios. Questions to be asked include: • Will the enterprise ever detect that the risk scenario has materialised? • Will the enterprise notice something has gone wrong so it can react appropriately? Generating scenarios and creatively thinking of what can go wrong will automatically raise and, hopefully, cause response to, the question of detectability. Detectability of scenarios includes two steps: visibility and recognition. The enterprise must be in a position that it can observe anything going wrong, and it needs the capability to recognise an observed event as something wrong.
Source: ISACA, *COBIT 5 for Risk*, USA, 2013, figure 37	

Cooperation with business continuity teams may be valuable because they are often working from the same type of scenarios, and avoiding unnecessary duplication of effort is always good practice. A common set of approaches to developing risk scenarios is to study past events, review audit reports, talk to users or managers, interview IT personnel (because they may know of common failures or unpatched vulnerabilities), and observe the business process or operations.

1.8.3 Benefits of Using Risk Scenarios

Risk scenarios provide a tool to facilitate communication in risk management. They allow for a narrative to be constructed that can communicate a relatable story to inspire stakeholders to take action.

The use of risk scenarios can enhance the risk management effort by helping the risk team to understand and explain risk to the business process owners and other stakeholders. A scenario provides a realistic and practical view of risk that is more aligned with business objectives, historical events and emerging threats. The risk scenario provides valuable information to the subsequent steps in the risk management process.

1.9 OWNERSHIP AND ACCOUNTABILITY

Risk requires ownership and accountability. After a risk has been identified, it is necessary to identify the owner of the risk. This is the manager or senior official in the organization who will bear the responsibility for determining the risk response, based on the risk acceptance level of the organization and the cost-benefit analysis of available controls or countermeasures. The owner is the person who will determine the level of control required to provide assurance of the effective management of the risk.

1.9.1 Principles of Risk and Control Ownership

Each risk should have an owner, and that owner should determine the controls that are necessary to mitigate that risk. Types of controls will be examined in more detail in chapter 3, Risk Response and Mitigation. The concept of a direct link between risk and control is important to ensure that all risk has been addressed through appropriate controls and that all controls are justified by the risk that mandates the requirement for that control. This is shown in **exhibit 1.23**.

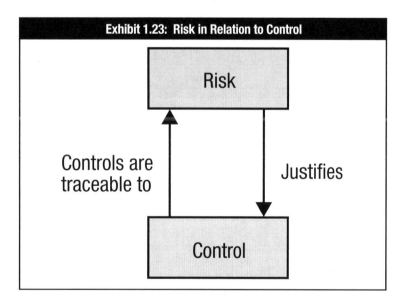

Exhibit 1.23: Risk in Relation to Control

A control is the means of managing risk, including policies, procedures, guidelines, practices or organizational structures, which can be of administrative, technical, management or legal nature. The implementation of a control results in a limitation and a cost to the organization. Even though a control can enable the use of new technology or a more efficient business process, it still presents a cost for implementation, maintenance and operation. A control forms part of the boundaries (voluntary or mandatory) in the control environment that shapes, guides and governs organizational and individual behavior and performance. It is crucial that the balance between performance and conformance be optimized for value creation and the promotion and protection of trust in the organization. A control may result in lower levels of system or network performance, or administrative or licensing costs. It may pose a

new risk to the organization if a failure of the control could lead to a denial of service. Therefore, the control should be justifiable, which can be proven by linking it back to a known risk or control objective. The presence of the risk justifies the selection and implementation of the control, and the performance of the control should be directly traceable back to the risk that the control is meant to mitigate. A control where there is no risk is not justifiable, and a risk without the enactment of appropriate, effective controls is a potential liability.

The owner of the risk, therefore, also owns the control and is responsible for monitoring its effectiveness. The control may be managed by the physical security department, a local manager, the IT department or other entity as a custodian on behalf of the owner, but the custodian is responsible to report on the status of the control to the risk and control owner. The control owner should be advised of any change in the risk environment or the effectiveness of the control so that adjustments to the control may be considered.

In some areas where there are regulations or laws that apply to risk, the risk owner may have to prepare standard reports on the status of risk, any incidents that may have occurred and the level of risk currently faced by the organization.

1.10 THE IT RISK REGISTER

The purpose of a risk register is to consolidate all information about risk into one central repository. This will allow management and each department to use a single resource to obtain the status of the risk management process. By documenting all risk in one location, it is possible to better manage and report on risk and coordinate risk response activities.

1.10.1 The Elements of a Risk Register

The risk register is a simple listing of all known risk. An example of a risk register template is shown in **exhibit 1.24**.

Exhibit 1.24: Example of a Risk Register Template			
Part I—Summary Data			
Risk statement			
Risk owner			
Date of last risk assessment			
Due date for update of risk assessment			
Risk category	☐ STRATEGIC (IT Benefit/Value Enablement)	☐ PROJECT DELIVERY (IT Programme and Project Delivery)	☐ OPERATIONAL (IT Operations and Service Delivery)
Risk classification (copied from risk analysis results)	☐ LOW	☐ MEDIUM	☐ HIGH ☐ VERY HIGH
Risk response	☐ ACCEPT	☐ TRANSFER	☐ MITIGATE ☐ AVOID
Part II—Risk Description			
Title			
High-level scenario (from list of sample high-level scenarios)			
Detailed scenario description—Scenario components	Actor		
	Threat Type		
	Event		
	Asset/Resource		
	Timing		
Other scenario information			

Net Producer Score

Exhibit 1.24: Example of a Risk Register Template *(cont.)*						
Part III—Risk Analysis Results						
Frequency of scenario (number of times per year)	0	1	2	3	4	5
	N ≤ 0,01 ☐	0,01 < N ≤ 0,1 ☐	0,1 < N ≤ 1 ☐	1 < N ≤ 10 ☐	10 < N ≤ 100 ☐	100 < N ☐
Comments on frequency						
Impact of scenario on business	0	1	2	3	4	5
1. Productivity	Revenue Loss Over One Year					
Impact rating	I ≤ 0,1% ☐	0,1% < I ≤ 1% ☐	1% < I ≤ 3% ☐	3% < I ≤ 5% ☐	5% < I ≤ 10% ☐	10% < I ☐
Detailed description of impact						
2. Cost of response	Expenses Associated With Managing the Loss Event					
Impact rating	I ≤ 10k$ ☐	10K$ < I ≤ 100K$ ☐	100K$ < I ≤ 1M$ ☐	1M$% < I ≤ 10M$ ☐	10M$ < I ≤ 100M$ ☐	100M$ < I ☐
Detailed description of impact						
3. Competitive advantage	Drop-in Customer Satisfaction Ratings					
Impact rating	I ≤ 0,5 ☐	0,5 < I ≤ 1 ☐	1 < I ≤ 1,5 ☐	1,5 < I ≤ 2 ☐	2 < I ≤ 2,5 ☐	2,5 < I ☐
Detailed description of impact						
4. Legal	Regulatory Compliance—Fines					
Impact rating	None ☐	< 1M$ ☐	< 10M$ ☐	< 100M$ ☐	< 1B$ ☐	> 1B$ ☐
Detailed description of impact						
Overall impact rating (average of four impact ratings)						

Overall rating of risk (obtained by combining frequency and impact ratings on risk map)	☐ LOW	☐ MEDIUM	☐ HIGH	☐ VERY HIGH

Part III—Risk Response				
Risk response for this risk	☐ ACCEPT	☐ TRANSFER	☐ MITIGATE	☐ AVOID
Justification				

Detailed description of response (NOT in case of ACCEPT)	Response Action	Completed	Action Plan
	1.	☐	☐
	2.	☐	☐
	3.	☐	☐
	4.	☐	☐
	5.	☐	☐
	6.	☐	☐
Overall status of risk action plan			
Major issues with risk action plan			
Overall status of completed responses			
Major issues with completed responses			

Part IV—Risk Indicators	
Key risk indicators for this risk	1. 2. 3. 4.

Source: ISACA, *COBIT 5 for Risk*, USA, 2013, figure 62

The risk register shows the severity, source and potential impact of a risk, as well as identifying the risk owner and the current status and disposition of the risk. The purpose of a risk register is to consolidate risk data into one place and permit the tracking of risk. The risk register is the one document that contains all risk that has been detected by various parts and activities of the organization including risk identified in audits, vulnerability assessments, penetration tests, incident reports, process reviews, management input, risk scenario creation and security assessments. This allows management to refer to one place to gain insight into the outstanding risk issues, the status of risk mitigation efforts and the emergence of newly identified and documented risk.

1.11 RISK CAPACITY, RISK APPETITE AND RISK TOLERANCE

Ownership of risk is ultimately the responsibility of the owners of the assets, or in most cases, senior management. Acceptance of risk is a formal and explicit statement of acknowledgement of risk and the conscious acceptance of the identified risk by the risk owner or responsible senior manager.

Risk acceptance must not exceed the risk capacity of the organization. Risk capacity is defined as the objective amount of loss an enterprise can tolerate without risking its continued existence. As such, it differs from risk appetite, which is a board/management decision on how much risk is desirable.

COBIT 5 for Risk notes that risk capacity and risk appetite are defined by board and executive management at the enterprise level. Some of the benefits of this approach include:
• Supporting and providing evidence of the risk-based decision-making processes
• Supporting the understanding of how each component of the enterprise contributes to the overall risk profile
• Showing how different resource allocation strategies can add to or lessen the burden of risk by simulating different risk response options
• Supporting the prioritization and approval process of risk response actions through risk budgets
• Identifying specific areas where a risk response should be made

Risk appetite is translated into a number of standards and policies to contain the risk level within the boundaries set by the risk appetite.[15] These boundaries need to be regularly adjusted or confirmed.

Risk tolerance levels are defined as tolerable deviations from the level set by the risk appetite definitions. Risk tolerance can be defined using IT process metrics or adherence to defined IT procedures and policies, which are a translation of the IT goals that need to be achieved. It coincides with risk appetite: risk tolerance is defined at the enterprise level and reflected in the policies created by senior management. Exceptions can be tolerated at lower levels of the enterprise as long as the overall exposure does not exceed the risk appetite at the enterprise level.[16]

Risk appetite and tolerance should be defined and approved by senior management and clearly communicated to all stakeholders, and a process should be in place to review and approve any exceptions.

As with all risk, risk appetite and tolerance change over time, and several factors (such as new technology, organizational structures, business strategy, etc.) require the organization to reassess its risk portfolio and reconfirm its risk appetite.[17]

NOTE: Risk tolerance is defined as the acceptable level of variation that management is willing to allow for any particular risk as the enterprise pursues its objectives. The interpretation of the ISACA definition is that while management has an official acceptance level of one value, they may accept a slight deviation from that level. An example of tolerance is a situation where the speed limit on a highway is 65 miles/hour, but a police officer may allow a person to travel at a speed up to 70 miles/hour before issuing a ticket.

Creating a risk-aware culture is an important aspect of monitoring risk tolerances. This may be a charge of senior management through revisiting and reinforcing the risk appetite and holding the proper stakeholders accountable for implementing risk management with the organization's risk appetite.

According to COSO, this is best accomplished when the organization communicates the risk appetite and risk tolerance, as discussed previously, with the following outcomes:[18]
• Consistent implementation across units
• Effective monitoring and communication of risk and changes in risk appetite
• Consistent understanding of risk appetite and related tolerances for each organizational unit
• Consistency between risk appetite, objectives, and relevant reward systems

1.11.1 Risk Acceptance
The risk practitioner should seek to know the risk appetite or acceptance level of the senior management team on behalf of the organization.

Some of the criteria for risk acceptance include:
• Consideration of the availability of mitigating controls
• The needs of regulation
• The cost-benefit analysis of a control option
• The risk-versus-reward incentive that management is willing to consider

Part of the challenge is the variation in risk appetite between different departments or lines of business. Some areas may be very risk averse and unwilling to take any risk unless absolutely necessary; however, other departments may be encouraged and expected to accept new challenges, take risk, try out new ideas and experiment. This variation in risk appetite can make the determination of appropriate risk treatment options quite challenging and requires careful documentation of the risk acceptance criteria.

For the most part, risk acceptance ensures that residual risk (the risk that remains after the implementation of risk treatment controls) is explicitly accepted by a level of management that is authorized to accept risk on behalf of the organization. When a risk that would exceed normal levels of residual risk has been accepted, management should document and sign off on the reasons for the acceptance decision.

ISO/IEC 31000 uses the term "tolerable level of risk" as synonymous with risk acceptance.

1.12 RISK AWARENESS
Awareness is a powerful tool in creating the culture, forming ethics and influencing the behavior of the members of an organization. The staff and operational teams of an organization are often the first to be aware of any problems or abnormal activities. Through awareness programs, it is possible to develop a team approach to risk management that enables every member of an organization to identify and report on risk and to work to defend systems and networks from attacks. Each member of the team can help identify vulnerabilities, suspicious activity and possible attacks. This may enable a faster response and better containment of a risk when an attack happens.

Risk awareness acknowledges that risk is an integral part of the business. This does not imply that all risk is to be avoided or eliminated, but rather that:
• Risk is well understood and known.
• IT risk issues are identifiable.
• The enterprise recognizes and uses the means to manage risk.

1.12.1 The Risk Awareness Program

A risk awareness program creates an understanding of risk, risk factors and the various types of risk that an organization faces. An awareness program should be tailored to the needs of the individual groups within an organization and deliver content suitable for that group. A risk awareness program should not disclose vulnerabilities or ongoing investigations except where the problem has already been addressed. Using examples and descriptions of the types of attacks and compromises that other organizations have experienced can reinforce the need for diligence and caution when addressing risk.

Awareness education and training can serve to mitigate some of the biggest organizational risk and achieve the most cost-effective improvement in risk and security. This can generally be achieved by educating an organization's staff in required procedures and policy compliance, as well as ensuring that staff can identify and understand the risk that threatens the organization. It is critical that the training effectively communicate the risk and its potential impact in order for staff to understand the justification for what many see as inconvenient extra steps that risk mitigation and security controls often require.

The risk practitioner must also understand the organization's structure and culture, as well as the types of communication that are most effective, to develop awareness and training programs that will be effective in the environment. Periodically changing risk awareness messages and the means of delivery will help maintain a higher level of risk awareness. Procedural controls can be complex, and it is essential to provide training as needed to ensure that staff understand the procedures and can correctly perform the required steps.

Awareness of information security policies, standards and procedures by all personnel is essential to achieving effective risk management. Employees cannot be expected to comply with policies or standards that they are not aware of, or follow procedures they do not understand. The risk practitioner should advise for a standardized approach, such as short computer- or paper-based quizzes to gauge awareness levels. Periodic use of a standardized testing approach provides metrics for awareness trends and training effectiveness. Further training needs can be determined by a skills assessment or employing a testing approach. Indicators for additional training requirements can come from various sources such as tracking help desk activity, operational errors, security events and audits.

An awareness program for management should highlight the need for management to play a supervisory role in protecting systems and applications from attack. A manager has the responsibility to oversee the actions of his/her staff and direct compliance with the policies and practices of the organization.

Awareness training for senior management should highlight the liability, need for compliance, due care and due diligence, and the need to create the tone and culture of the organization through policy and good practice. Senior management may need to be reminded that they are the ones who "own" the risk and bear the responsibility for determining risk acceptance levels.

1.13 SUMMARY

This chapter addresses the initial phase of a risk management effort. The intention of risk identification is to set out a clear path for the later chapters of risk assessment, risk response, and risk and control monitoring.

The core learning objectives of this chapter are to identify risk factors such as assets, threats, vulnerabilities, consequence and likelihood. These are documented and used in the next steps of the risk management process to assess the risk and ultimately drive the selection of appropriate controls or another suitable risk response.

ENDNOTES

1 International Organization for Standardization (ISO); *ISO 31000:2009: Risk Management—Principles and guidelines*, Switzerland, 2009

2 ISACA, *COBIT® 5 for Risk*, USA, 2013, *www.isaca.org/cobit*

3 ISO; *IEC 31010:2009: Risk Management—Risk Assessment Techniques*, Switzerland, 2009

4 ISO/International Electrotechnical Commission (IEC); *ISO/IEC 27001:2013: Information Technology – Security Techniques—Information Security Management Systems—Requirements*, Switzerland, 2013

5 ISO/IEC; *ISO/IEC 27005:2011: Information Technology—Security Techniques—Information Security Risk Management*, Switzerland, 2011

6 National Institute of Standards and Technology (NIST); *NIST Special Publication 800-30 Revision 1: Guide for Conducting Risk Assessments*, USA, 2012

7 NIST; *NIST Special Publication 800-39: Managing Information Security Risk*, USA, 2011

8 European Union, *Regulation EC No 45/2001: Data Protection by Community Institutions and Bodies*, *europa.eu/legislation_summaries/information_society/data_protection/l24222_en.htm*

9 *Op cit* ISO/IEC 27005

10 Verizon; *2014 Data Breach Investigations Report*, USA, 2014, *www.verizonenterprise.com/DBIR/2014*

11 Verizon; *2013 Data Breach Investigations Report*, USA, 2013, *www.verizonenterprise.com/DBIR/2013/series*

12 The Open Web Application Security Project (OWASP), *www.owasp.org*

13 ISACA, *Generating Value from Big Data Analytics*, USA, 2014, *www.isaca.org/Knowledge-Center/Research/Documents/Generating-Value-from-Big-Data-Analytics_whp_Eng_0114.pdf*

14 *Op cit* ISACA, 2013

15 *Ibid.*

16 *Ibid.*

17 *Ibid.*

18 Rittenburg, Larry; Frank Martens; *Enterprise Risk Management – Understanding and Communicating Risk Appetite*, Committee of Sponsoring Organizations of the Treadway Commission (COSO), USA, 2012

Page intentionally left blank

Chapter 2: IT Risk Assessment

Section One: Overview

Section Two: Contents

Section One: Overview

DOMAIN DEFINITION

Analyze and evaluate IT risk to determine the likelihood and impact on business objectives to enable risk-based decision making.

LEARNING OBJECTIVES

The objective of this domain is to ensure that the CRISC candidate has the knowledge necessary to:
• Identify and apply risk assessment techniques
• Analyze risk scenarios
• Identify current state of controls
• Assess gaps between current and desired states of the IT risk environment
• Communicate IT risk assessment results to relevant stakeholders.

CRISC EXAM REFERENCE

This domain represents 28 percent of the CRISC exam (approximately 42 questions).

TASK AND KNOWLEDGE STATEMENTS

TASKS

There are six tasks within this domain that a CRISC candidate must know how to perform. These relate to the IT risk assessment process.

T2.1   Analyze risk scenarios based on organizational criteria (e.g., organizational structure, policies, standards, technology, architecture, controls) to determine the likelihood and impact of an identified risk.
T2.2   Identify the current state of existing controls and evaluate their effectiveness for IT risk mitigation.
T2.3   Review the results of risk and control analysis to assess any gaps between current and desired states of the IT risk environment.
T2.4   Ensure that risk ownership is assigned at the appropriate level to establish clear lines of accountability.
T2.5   Communicate the results of risk assessments to senior management and appropriate stakeholders to enable risk-based decision making.
T2.6   Update the risk register with the results of the risk assessment.

KNOWLEDGE STATEMENTS

The CRISC candidate should be familiar with the task statements relevant to each domain in the CRISC job practice. The tasks are supported by 41 knowledge statements that delineate each of the areas in which the risk practitioner must have a good understanding in order to perform the tasks. Many knowledge statements support tasks that cross domains.

The CRISC candidate should have knowledge of:
1. Laws, regulations, standards and compliance requirements
2. Industry trends and emerging technologies
3. Enterprise systems architecture (e.g., platforms, networks, applications, databases and operating systems)
4. Business goals and objectives
5. Contractual requirements with customers and third-party service providers
6. Threats and vulnerabilities related to:
    6.1. Business processes and initiatives
    6.2. Third-party management
    6.3. Data management
    6.4. Hardware, software and appliances
    6.5. The system development life cycle (SDLC)

Chapter 2: IT Risk Assessment

Section One: Overview

Certified in Risk
and Information
Systems Control
An ISACA Certification
CRISC

    6.6. Project and program management

    6.7. Business continuity and disaster recovery management (DRM)

    6.8. Management of IT operations

    6.9. Emerging technologies

7. Methods to identify risk

8. Risk scenario development tools and techniques

9. Risk identification and classification standards, and frameworks

10. Risk events/incident concepts (e.g., contributing conditions, lessons learned, loss result)

11. Elements of a risk register

12. Risk appetite and tolerance

13. Risk analysis methodologies (quantitative and qualitative)

14. Organizational structures

15. Organizational culture, ethics and behavior

16. Organizational assets (e.g., people, technology, data, trademarks, intellectual property) and business processes, including enterprise risk management (ERM)

17. Organizational policies and standards

18. Business process review tools and techniques

19. Analysis techniques (e.g., root cause, gap, cost-benefit, return on investment [ROI])

20. Capability assessment models and improvement techniques and strategies

21. Data analysis, validation and aggregation techniques (e.g., trend analysis, modeling)

22. Data collection and extraction tools and techniques

23. Principles of risk and control ownership

24. Characteristics of inherent and residual risk

25. Exception management practices

26. Risk assessment standards, frameworks and techniques

27. Risk response options (i.e., accept, mitigate, avoid, transfer) and criteria for selection

28. Information security concepts and principles, including confidentiality, integrity and availability of information

29. Systems control design and implementation, including testing methodologies and practices

30. The impact of emerging technologies on design and implementation of controls

31. Requirements, principles, and practices for educating and training on risk and control activities

32. Key risk indicators (KRIs)

33. Risk monitoring standards and frameworks

34. Risk monitoring tools and techniques

35. Risk reporting tools and techniques

36. IT risk management best practices

37. Key performance indicator (KPIs)

38. Control types, standards, and frameworks

39. Control monitoring and reporting tools and techniques

40. Control assessment types (e.g., self-assessments, audits, vulnerability assessments, penetration tests, third-party assurance)

41. Control activities, objectives, practices and metrics related to:

    41.1. Business processes

    41.2. Information security, including technology certification and accreditation practices

    41.3. Third-party management, including service delivery

    41.4. Data management

    41.5. The system development life cycle (SDLC)

    41.6. Project and program management

    41.7. Business continuity and disaster recovery management (DRM)

    41.8. IT operations management

    41.9. The information systems architecture (e.g., platforms, networks, applications, databases and operating systems)

SELF-ASSESSMENT QUESTIONS

2-1   The **MOST** significant drawback of using quantitative risk analysis instead of qualitative risk analysis is the:

    A. lower objectivity.
    B. higher reliance on skilled personnel.
    C. lower management buy-in.
    D. higher cost.

2-2   Risk scenarios are analyzed to determine:

    A. strength of controls.
    B. likelihood and impact.
    C. current risk profile.
    D. scenario root cause.

2-3   The risk to an information system that supports a critical business process is owned by:

    A. the IT director.
    B. senior management.
    C. the risk management department.
    D. the system users.

2-4   The **PRIMARY** reason risk assessments should be repeated at regular intervals is:

    A. omissions in earlier assessments can be addressed.
    B. periodic assessments allow various methodologies.
    C. business threats are constantly changing.
    D. they help raise risk awareness among staff.

2-5   Which of the following choices **BEST** assists a risk practitioner in measuring the existing level of development of risk management processes against their desired state?

    A. A capability maturity model (CMM)
    B. Risk management audit reports
    C. A balanced scorecard (BSC)
    D. Enterprise security architecture

2-6   Which of the following choices **BEST** helps identify information systems control deficiencies?

    A. Gap analysis
    B. The current IT risk profile
    C. The IT controls framework
    D. Countermeasure analysis

2-7   Deriving the likelihood and impact of risk scenarios through statistical methods is **BEST** described as:

A. risk scenario analysis.
B. qualitative risk analysis.
C. quantitative risk analysis.
D. probabilistic risk assessment.

2-8   Which of the following reviews is **BEST** suited for the review of IT risk analysis results before the results are sent to management for approval and use in decision making?

A. An internal audit review
B. A peer review
C. A compliance review
D. A risk policy review

Malware — pharmis attack

IT Related Risk Scenarios — Business Objectives

ANSWERS TO SELF-ASSESSMENT QUESTIONS

Correct answers are shown in **bold**.

2-1   A.   Neither of the two risk analysis methods is fully objective. While the qualitative method subjectively assigns high, medium and low frequency and impact categories to a specific risk, subjectivity within the quantitative method is often expressed in mathematical "weights."
       B.   To be effective, both processes require personnel who have a good understanding of the business.
       C.   Quantitative analysis generally has a better buy-in than qualitative analysis to the point where it can cause overreliance on the results.
       D.   Quantitative risk analysis is generally more complex and, therefore, more costly than qualitative risk analysis.

2-2   A.   The strength of controls is determined after the controls are in place to ensure they are adequate in addressing the risk.
       B.   Risk scenarios are descriptions of events that can lead to a business impact and are evaluated to determine the likelihood and impact should the event occur.
       C.   The current risk profile is the identification of risk currently of concern by the organization.
       D.   The risk scenario process is used to identify plausible scenarios and from there determine likelihood and impact. Determining a root cause is not a part of the risk scenario process.

2-3   A.   The IT director manages the IT systems on behalf of the business owners.
       B.   Senior management is responsible for the acceptance and mitigation of all risk.
       C.   The risk management department determines and reports on level of risk, but does not own the risk.
       D.   The system users are responsible for utilizing the system properly and following procedures, but they do not own the risk.

2-4   A.   Performing risk assessments on a periodic basis can find omissions in earlier assessments, but this is not the primary reason for conducting regular reassessments.
       B.   Organizations strive to improve their risk management process to more quickly and accurately assess and address risk, and this may involve changing the methodology. However, it is not the primary reason for conducting regular assessments.
       C.   As business objectives and methods change, the nature and relevance of threats also change. This is the primary reason to conduct periodic risk assessments.
       D.   Risk assessments are conducted on a periodic basis to address new threats and changes in the business. Creating more risk awareness is a minor benefit of conducting periodic risk assessments.

2-5   **A.   The capability maturity model (CMM) grades processes on a scale of 0 to 5, based on their maturity, and is commonly used by entities to measure their existing state and then to determine the desired one.**
       B.   Risk management audit reports offer a limited view of the current state of risk management.
       C.   A balanced scorecard (BSC) enables management to measure the implementation of strategy and assists in its translation into action.
       D.   Enterprise security architecture explains the security architecture of an entity in terms of business strategy, objectives, relationships, risk, constraints and enablers and provides a business-driven and business-focused view of security architecture.

2-6   A.   **Controls are deployed to achieve the desired control objectives based on risk assessments and business requirements. The gap between desired control objectives and actual IS control design and operational effectiveness identifies IS control deficiencies.**

       B.   Without knowing the gap between desired state and current state, one cannot identify the control deficiencies.

       C.   The IT controls framework is a generic document with no information such as desired state of IS controls and current state of the enterprise; therefore, it will not help in identifying IS control deficiencies.

       D.   Countermeasure analysis only helps in identifying deficiencies in countermeasures, not in the full set of primary controls.

2-7   A.   A risk scenario analysis generally includes several risk analysis methods, including quantitative, semiquantitative and qualitative. The question stem describes only the quantitative risk analysis method.

       B.   A qualitative risk analysis would use nonquantitative measures to estimate the likelihood and impact of adverse events; these might include low, medium and high for likelihood level and low, medium, high and catastrophic for impact level.

       C.   **The essence of quantitative risk assessment is to derive the likelihood and impact of risk scenarios based on statistical methods and data.**

       D.   Probabilistic risk assessments are mostly used for the assessment of risk related to complex engineered technology (e.g., nuclear plants, airplanes). They rely on a systematic and comprehensive methodology and consider both quantitative and qualitative risk analysis. The question stem describes only the quantitative risk analysis method.

2-8   A.   An internal audit review is not best suited for the review of IT risk analysis results. Internal auditing is an independent, objective assurance and consulting activity designed to add value and improve an enterprise's operations. It helps an organization accomplish its objectives by bringing a systematic, disciplined approach to evaluate and improve the effectiveness of risk management, control and governance processes.

       B.   **It is effective, efficient and good practice to perform a peer review of IT risk analysis results before sending them to management.**

       C.   A compliance review is not best suited for the review of IT risk analysis results. Compliance reviews measure the conformance with a specific, measurable standard.

       D.   A review of the risk policy will change the contents and methods of the risk analysis eventually, but this is not a way of reviewing IT risk analysis results before sending them to management.

NOTE: For more self-assessment questions, you may also want to obtain a copy of the *CRISC™ Review Questions, Answers & Explanations Manual 2015*, which consists of 400 multiple-choice study questions, answers and explanations, and the *CRISC™ Review Questions, Answers & Explanations Manual 2015 Supplement*, which consists of 100 new multiple-choice study questions, answers and explanations.

SUGGESTED RESOURCES FOR FURTHER STUDY

In addition to the resources cited throughout this manual, the following resources are suggested for further study in this domain (publications in **bold** are stocked in the ISACA Bookstore):

Committee of Sponsoring Organizations of the Treadway Commission (COSO); *Enterprise Risk Management for Cloud Computing*, USA, 2012

International Organization for Standardization (ISO); *IEC 31010:2009: Risk management—Risk assessment techniques*, Switzerland, 2009

ISO/International Electrotechnical Commission (IEC); *ISO/IEC 27005:2008: Information technology—Security techniques—Information security risk management*, Switzerland, 2008

ISACA, COBIT® 5, USA, 2012, *www.isaca.org/cobit*

ISACA, *COBIT® 5 for Risk*, **USA, 2013,** *www.isaca.org/cobit*

ISACA, *The Risk IT Framework*, **USA, 2009**

ISACA, *The Risk IT Practitioner Guide*, **USA, 2009**

Kissel, Richard; Kevin Stine; Matthew Scholl; Hart Rossman; Jim Fahlsing; Jessica Gulick; *NIST Special Publication 800-64 Revision 2: System Considerations in the System Development Life Cycle*, National Institute of Standards and Technology (NIST), USA, 2008

National Institute of Standards and Technology (NIST); *NIST Special Publication 800-30 Revision 1: Guide for Conducting Risk Assessments*, USA, 2012

Panda, Parthajit, "The OCTAVE Approach to Information Security Risk Assessment," *ISACA Journal*, Volume 4, 2009

Section Two: Content

2.0 OVERVIEW

The processes described in chapter 1, IT Risk Identification, revealed the risk facing the organization. Risk factors were identified and documented in the risk register. The next step in the risk management life cycle, as shown in **exhibit 2.1**, takes the foundational work performed previously and assesses the IT risk level. IT risk is a subset of enterprise risk. The risk faced by an IT system is most often measured by the impact of an IT-related problem on the business services that the IT system supports. Therefore, the calculation or assessment of IT's impact must consider the dependencies of the other systems, departments, business partners and users on the affected IT system.

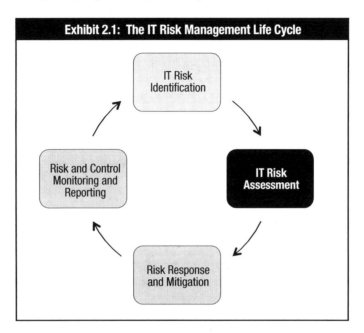

Exhibit 2.1: The IT Risk Management Life Cycle

The calculation of the risk assessment is essential to provide the information required in chapter 3, Risk Response and Mitigation. The choice of an appropriate response is dependent on the accuracy of the data provided from the IT risk assessment effort.

2.1 RISK IDENTIFICATION VERSUS RISK ASSESSMENT

Risk identification is the process of determining and documenting the risk that an enterprise faces. The identification of risk is based on the recognition of threats, vulnerabilities, assets and controls in the enterprise's operational environment.

Risk identification begins with documenting the assets of the organization and determining the value of an asset. This is measured in several ways, including the impact of that asset in meeting business goals and profitability, the relationship between the asset and reputation, the cost to recover or rebuild the asset if lost, the liabilities or fines for asset compromise, and the value of the asset to a competitor.

Risk identification also includes documenting threats that could pose a risk of damage to the organization. This is based on historical data; available resources to list the threats; and the observation and examination of the business, the culture of the organization and the persistence of the adversary. This step also includes examining any gaps or weaknesses in the organization's information security and documenting the state of existing controls.

Risk assessment is defined as a process used to identify and evaluate risk and its potential effects. Risk assessment includes assessing the critical functions necessary for an enterprise to continue business operations, defining the controls in place to reduce exposure and evaluating the cost for such controls. Risk analysis often involves an evaluation of the probabilities of a particular event.

There are several well-known risk management methodologies and standards; however, they describe the relationships between risk identification and risk assessment differently. Some standards specifically state that risk identification is a component of risk assessment, whereas other standards describe the two as separate processes.

IEC 31010:2009 states:[1]
>   *Risk assessment attempts to answer the following fundamental questions:*
>   • *What can happen and why (by risk identification)?*
>   • *What are the consequences?*
>   • *What is the probability of their future occurrence?*
>   • *Are there any factors that mitigate the consequence of the risk or that reduce the probability of the risk?*

ISO/IEC 27005 describes risk assessment as follows:[2]
>   *Risk assessment consists of the following activities:*
>   • *Risk analysis which comprises:*
>   – *Risk identification*
>   – *Risk estimation*
>   – *Risk evaluation*

NOTE: The CRISC candidate will not be tested on any specific standard. The use of any standards in this review manual is for example and explanatory purposes only.

2.2 RISK ASSESSMENT TECHNIQUES

There are several approaches to risk analysis and assessment. Each approach has certain advantages and possible weaknesses. Using a consistent risk assessment methodology or framework is more important than which one is used.

A list of risk assessment techniques is shown in **exhibit 2.2**.

Exhibit 2.2: Risk Assessment Techniques
Bayesian statistics and Bayes nets
Bow tie analysis
Brainstorming/Structured or semistructured interviews
Business impact analysis (BIA)
Cause and consequence analysis
Cause-and-effect analysis
Checklists
Delphi method
Environmental risk assessment
Event tree analysis
Fault tree analysis
Hazard analysis and critical control points (HACCP)
Hazard and operability study (HAZOP)
Human reliability analysis (HRA)
Layers of protection analysis (LOPA)
Markov analysis
Monte Carlo simulation

Exhibit 2.2: Risk Assessment Techniques *(cont.)*
Preliminary hazard analysis
Reliability-centered maintenance
Root cause analysis (pre-mortems)
Scenario analysis
Sneak circuit analysis
Structured "what if" technique (SWIFT)
Adapted from *IEC 31010:2009 Risk management—Risk assessment techniques,* Switzerland, 2009

2.2.1 Bayesian Analysis

A Bayesian analysis is a method of statistical inference that uses prior distribution data to determine the probability of a result. This technique relies on the prior distribution data to be accurate to be effective and to produce accurate results.

2.2.2 Bow Tie Analysis

A bow tie analysis provides a diagram to communicate risk assessment results by displaying links between possible causes, controls and consequences. The cause of the event is depicted in the middle of the diagram (the "knot" of the bow tie) and triggers, controls, mitigation strategies and consequences branch off of the "knot."

2.2.3 Brainstorming/Structured Interview

The purpose of the structured interview and brainstorming is to gather a large group of types of potential risk or ideas and have a team rank them. The initial interview or brainstorming may be completed using prompts or interviews with an individual or small group. For more information on interviewing techniques, see chapter 1, section 1.2.1.

2.2.4 Business Impact Analysis

A BIA is a process to determine the impact of losing the support of any resource. The BIA will establish the escalation of that loss over time. It is predicated on the fact that senior management, when provided reliable data to document the potential impact of a lost resource, can make the appropriate decision.

2.2.5 Cause and Consequence Analysis

A cause and consequence analysis combines techniques of a fault tree analysis and an event tree analysis and allows for time delays to be considered.

2.2.6 Cause-and-effect Analysis

A cause-and-effect analysis looks at the factors that contributed to a certain effect and groups the causes into categories (using brainstorming), which are then displayed using a diagram, typically a tree structure or a fishbone diagram.

2.2.7 Checklists

A checklist is a list of potential or typical threats or other considerations that should be of interest to the organization. The risk practitioner will use a previously developed list, codes or standards to assess the risk using this method.

2.2.8 Delphi Method

The Delphi method uses expert opinion, which is often received using two or more rounds of questionnaires. After each round of questioning, the results are summarized and communicated to the experts by a facilitator. This collaborative technique is often used to build a consensus among the experts.

2.2.9 Event Tree Analysis

An event tree analysis is a forward, bottom up model that uses inductive reasoning to assess the probability of different events resulting in possible outcomes.

2.2.10 Fault Tree Analysis

A fault tree analysis starts with an event and examines possible means for the event to occur (top down) and displays these results in a logical tree diagram. This diagram can be used to generate ways to reduce or eliminate potential causes of the event.

2.2.11 Hazard Analysis and Critical Control Points (HACCP)

Originally developed for the food safety industry, HACCP is a system for proactively preventing risk and assuring quality, reliability and safety of processes. The system monitors specific characteristics, which should fall within defined limits.

2.2.12 Hazard and Operability Studies (HAZOP)

A HAZOP is a structured means of identifying and evaluating potential risk by looking at possible deviations from existing processes.

2.2.13 Human Reliability Analysis (HRA)

HRA examines the effect of human error on systems and their performance.

2.2.14 Layers of Protection Analysis (LOPA)

LOPA is a semiquantitative risk analysis technique that uses aspects of HAZOP data to determine risk associated with risk events. It also looks at controls and their effectiveness.

2.2.15 Markov Analysis

A Markov analysis is used to analyze systems that can exist in multiple states. The Markov model assumes that future events are independent of past events.

2.2.16 Monte-Carlo Analysis

IEC 31010:2009 describes Monte Carlo simulation as the following:[3]

> *Monte Carlo simulation is used to establish the aggregate variation in a system resulting from variations in the system, for a number of inputs, where each input has a defined distribution and the inputs are related to the output via defined relationships. The analysis can be used for a specific model where the interactions of the various inputs can be mathematically defined. The inputs can be based upon a variety of distribution types according to the nature of the uncertainty they are intended to represent. For risk assessment, triangular distributions or beta distributions are commonly used.*

2.2.17 Preliminary Hazard Analysis

Preliminary hazard analysis looks at what threats or hazards may harm an organization's activities, facilities or systems. The result is a list of potential risk.

2.2.18 Reliability-centered Maintenance

Reliability-centered maintenance analyzes the functions and potential failures of a specific asset, particularly a physical asset such as equipment.

2.2.19 Root Cause Analysis

Root cause analysis is discussed in further detail in section 2.5.3 in this chapter.

2.2.20 Scenario Analysis

Scenario analysis examines possible future scenarios that were identified during risk identification, looking for risk associated with the scenario should it occur. Section 2.3 in this chapter discusses this technique in more detail.

2.2.21 Sneak Circuit Analysis

A sneak circuit analysis is used to identify design errors or sneak conditions—latent hardware, software or integrated conditions that may cause an unwanted event to occur. They are often undetected by system tests. These can result in improper operations, loss of availability, program delays or injury to personnel.

2.2.22 Structured "What If" Technique (SWIFT)

A structured "what if" technique uses structured brainstorming to identify risk, typically within a facilitated workshop. It uses prompts and guide words and is typically used with another risk analysis and evaluation technique.

2.3 ANALYZING RISK SCENARIOS

During risk identification, risk scenarios are developed and used to identify and describe potential risk events. These scenarios are useful to communicate with the business and gather input data required to understand the potential or probable impact of the risk event if it were to occur.

The impact of a risk event is hard to calculate with any degree of accuracy because there are many factors that affect the outcome of an event. If the event is detected quickly and appropriate measures are taken to contain the incident, then the impact may be minimized and the recovery process may be fairly rapid. However, if the organization is unable to detect the incident promptly, the same incident could cause severe damage and result in much higher recovery costs. Some of the factors that can affect the calculation of risk assessment are discussed in the following sections.

2.3.1 Organizational Structure and Culture

The structure and culture of the organization are contributing factors in risk prevention, risk detection and risk response. A mature organization has policies and procedures and an effective reporting and notification structure in place to detect, notify and escalate a situation effectively. An organization that does not have a mature incident response capability will often react to incidents in an unpredictable, *ad hoc*, reactive manner and will experience inconsistent results.

The risk management function should have an enterprisewide mandate that allows the risk management team to review and provide input into all business processes. They should participate in the incident management activities and be responsible for investigating incidents to ensure that all lessons are learned and to improve incident response planning, detection and recovery. It is important to note that lessons learned in one department may be applicable in protecting other departments from the same problems.

If the culture of the organization is to hide problems rather than communicate or address them—or to only use an adverse situation to point blame—then the ability of the risk practitioner to effectively contribute to the protection of the organization and assist in the investigation of an incident may be severely impaired.

2.3.2 Policies

Policies provide direction regarding acceptable and unacceptable behaviors and actions to the organization and send a clear message from senior management regarding the desired approach to the protection of assets and the culture of the organization. Policies give authority to the staff of the risk management, audit and security teams of the organization to perform their job responsibilities. The policies of the organization should clearly state the position of senior management toward the protection of information. This will lead to the development of procedures, standards and baselines that implement the intent of policy and mandate that all departments comply with the requirements of policy.

There are often several layers of policies. A high-level policy is issued by senior management as a way to address the objectives of the organization's mission and vision statement. This policy, often called "the overarching security policy," is written at a high level and does not have a technical focus so that it does not become outdated as soon as technology changes. High-level policy should require compliance with laws and best practices and state the goal of managing risk through protecting the organization's assets, including the information and information systems that support business operations.

The next level of policies is technical and functional and includes specifics regarding the use of technology, i.e., a remote access policy, acceptable use policy, and password and access policy. These policies are subject to change as technology changes and new systems are developed.

High-level policies are instrumental in determining the approach of the organization toward risk management and acceptable levels of risk. Without policies in place, the risk practitioner may not be able to gain access to key personnel, be left out of strategic planning sessions and be ignored when performing investigations.

The risk practitioner should be concerned with the risk associated with the presence or lack of policies and whether the policies are enforced. If policies are not developed and communicated, the organization has no means of enforcing standards of behavior, which poses a risk of inappropriate behavior occurring. A lack of enforcement may also lead to the risk of people circumventing the controls and deciding that they do not have to comply with policy. A lack of enforcement may also lead to increased liability because the organization has admitted the need for a policy, but does not follow its own rules as mandated by policy.

The risk practitioner should assess the risk associated with the policy framework of the organization and provides recommendations as necessary. Where policies are out of date, unenforced or incomplete, the risk practitioner should underline the vulnerability and risk that this poses to the organization.

2.3.3 Standards and Procedures

Standards and procedures support the requirements defined in the policies set by the organization. A standard is defined as a mandatory requirement, code of practice or specification approved by a recognized external standards organization, such as the International Organization for Standardization (ISO). Standards are implemented to comply with the requirements and direction of policy to limit risk and support efficient business operations.

Many organizations realize that the value of standards is the authority and proven value that they provide. A standard mandates the way in which personnel in an organization must comply with recognized practices or specifications and can help ensure a consistent approach to meeting risk requirements across an organization. An organization may base its practices and operations on external standards such as an ISO standard, or it may develop its own standards, such as requiring all staff to use the same product, operating system or desktop. The use of a standard facilitates support and maintenance, provides better cost control, and provides authority for the practices and procedures of the organization because a standard requires the implementation of certain practices.

A procedure is a document containing a detailed description of the steps necessary to perform specific operations in conformance with applicable standards. Procedures are defined as part of processes.

Procedures are invaluable to implementing the intent of policy. They describe the consistent and measureable ways that an operation is conducted so that the risk practitioner can be assured that operations are performed properly and any abnormal operations can be detected.

A lack of standards and procedures will result in undependable, inconsistent operations and may result in risk due to not detecting a risk event, noncompliance with regulations or difficulty preventing an attack.

2.3.4 Technology

A considerable risk factor is the age and condition of technology in use in the organization. Many organizations use outdated technologies, which can be difficult to obtain, support and maintain. The technology in use may have been acquired through multiple projects and/or mergers. It may, therefore, consist of products from a varied mix of vendors, languages, configurations and vintages, or it may run on different operating systems, hardware, network architectures and databases. A large organization may have systems based on mainframes, client-servers, virtual environments and the cloud all at once. The complexity that this represents is a source of risk. The best practice is to have a simple, easily managed and controlled environment with standard products and technologies, but this may be nearly impossible to attain in a large organization. Many systems may continue to be used long after their anticipated life span simply because it is not cost-effective to replace them.

Some of the considerations that affect risk assessment related to technology include:
- Age of equipment
- Expertise available for maintenance
- Variety of vendors/suppliers (are they still in business?)
- Documentation of systems
- Availability of replacement parts
- Ability to test systems or equipment
- Operating environment and user expertise
- Ability to patch/mitigate vulnerabilities

2.3.5 Architecture

A key factor in the maturation of the processes and practices of an organization is the development of an enterprisewide approach to risk management, architecture and business continuity. The development of an enterprisewide approach will promote consistency, repeatability, compliance, accountability and visibility to senior management into the practices and strategy of the organization.

For example, very few organizations have a mature IT architecture. The systems used by most organizations have been built as part of individual projects or initiatives, and each system is an independent entity with little in common with other systems. The lack of architecture results in gaps between system ownership and unclear areas of responsibility for incidents or configuration management. The more complex and undefined an architecture is, the more challenging it is to secure the entire network and ensure compliance with security standards, regulations and good practices. It is not uncommon to find that the systems and networks of an organization have not been documented or mapped out on a network diagram. This means that the ability of the organization to measure risk or identify vulnerabilities is severely impacted.

A lack of architecture often results in:
• Controls that overlap
• Controls that conflict with one another
• Unidentified single points of failure
• Unidentified methods to bypass controls
• Inadequate network isolation

A lack of a defined, cohesive and documented architecture is a serious risk for the risk practitioner to consider when assessing risk. The most serious problem is that the organization may not be aware of many of their systems and network problems because they do not know what vulnerabilities they have.

2.3.6 Controls

When assessing risk, the risk practitioner must take into consideration the current control environment. Controls are implemented to mitigate risk or to comply with regulations; however, the risk practitioner may find that many controls are not working correctly, are poorly maintained, are not suitable in relation to the risk or are incorrectly configured.

A review of the controls evaluates whether the controls are working effectively to mitigate the risk and that there is the correct balance between technical, managerial (administrative), and physical or operational control types. Implementation of a technical control, such as a firewall, requires correct training for the staff who will manage or operate the control, correct procedures for configuring the control, assignment of responsibilities for monitoring the control and reviewing the log data generated by the firewall, and regular testing of the functions of the control. Implementing a technological control also requires adequate coinciding controls to ensure that the technology control is being effectively managed or monitored. If these coinciding controls are not in place, stakeholders may develop a false sense of security, and this could lead to a serious risk of unidentified vulnerabilities and/or an ineffective use of resources.

Controls can be categorized as compensating, corrective, detective, deterrent, directive and preventive. These categories are described in **exhibit 2.3**.

Exhibit 2.3: Control Categories	
Category	**Description**
Compensating controls	An alternate form of control that corrects a deficiency or weakness in the control structure of the enterprise Compensating controls may be considered when an entity cannot meet a requirement explicitly, as stated, due to legitimate technical or business constraints, but has sufficiently mitigated the risk associated with the requirement through implementation of other controls. **Example:** Adding a challenge response component to weak access controls can compensate for the deficiency in the access control mechanism.
Corrective controls	Remediate errors, omissions and unauthorized uses and intrusions, once they are detected **Example:** Backup restore procedures enable a system to be recovered if harm is so extensive that processing cannot continue without recourse to corrective measures.
Detective controls	Warn of violations or attempted violations of security policy and include such controls as audit trails, intrusion detection methods and checksums
Deterrent controls	Provide warnings that can deter potential compromise, such as warning banners on login screens or offering rewards for the arrest of hackers
Directive controls	Mandate the behavior of an entity by specifying what actions are, or are not, permitted; a directive control is often considered to be a type of deterrent. **Example:** A policy is an example of directive control.
Preventive controls	Inhibit attempts to violate security policy and include such controls as access control enforcement, encryption and authentication

Handwritten margin notes: "additional controls", "Store Procedure", "IDS/IPS log", "encryption authentication"

The interaction between the control types is shown in **exhibit 2.4**.

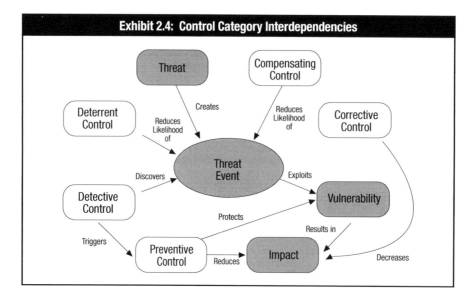

The risk practitioner must assess the current control environment to evaluate the risk culture and effectiveness of the current risk management program. This will indicate the level of risk currently facing the organization and the seriousness of risk. The risk is much more serious if:
• Controls are inadequate
• The wrong controls are being used
• Controls are ignored or bypassed
• Controls are poorly maintained
• Logs or control data are not reviewed
• Controls are not tested
• Changes to the configuration of controls are not managed
• Controls can be physically accessed and altered

> **The selection of controls is addressed in chapter 3, Risk Response and Mitigation, and the monitoring of controls is addressed in chapter 4, Risk and Control Monitoring and Reporting.**

2.4 CURRENT STATE OF CONTROLS

The risk practitioner uses regular reports generated by controls and the results of control testing activities, along with the results of incident management programs to determine the current state of IT risk. Current state refers to the condition of the program at a point in time, so the IT risk assessment is only valid at the time the risk state is measured. However, regular reviews of IT risk—as seen in chapter 4, Risk and Control Monitoring and Reporting—are used to determine the state of IT risk on a regular basis and with scheduled reporting to management.

For the purposes of IT risk assessment, the risk practitioner must review current risk levels and contrast the current level with the desired state or level of acceptable risk. A gap or disparity between the current and desired state must be investigated to determine the reason for the gap, and solutions should be recommended to address the disparity.

Some of the tools used by the risk practitioner to determine the current state of IT risk are:
• Audits
• Control tests conducted by the control owner or custodian
• Incident reports
• IT operations feedback
• Logs
• Media reports of new vulnerabilities and threats
• Observation
• Self-assessments
• Third-party assurance
• User feedback
• Vendor reports
• Vulnerability assessments and penetration tests

2.4.1 Audits

An audit is an excellent source of analysis data and recommendations for the risk practitioner. Recommendations from audits are often related to the improvement of managerial, technical and operational controls.

2.4.2 Control Tests

A control is selected to address one or more specific types of risk; however, the control, as designed, implemented and maintained, may not provide the desired level of protection or work effectively. The risk practitioner must test the performance of the control to ensure that it is properly installed, operating correctly and providing the desired result. The risk practitioner should advise the organization on the risk of a control being bypassed, disabled or rendered ineffective through changes to the operating environment, the configuration of the control or an emerging, previously unidentified threat.

Testing a control includes testing both the technical and nontechnical aspects of the control, such as the rules governing the operation of the control, the procedures used in monitoring and operating the control, and the proficiency of the staff responsible for the operation of the control.

2.4.3 Incident Reports
The risk practitioner may learn about threats, vulnerabilities and impact from the review of incident reports. A thorough review of an incident can identify weak controls, poor detection, inappropriate or ineffective response, and lack of training of staff.

2.4.4 IT Operations Feedback
When problems arise, IT operations staff is usually the first to hear about them; therefore, they are aware of the common problems a system may have. Interviewing operations staff and reviewing logs and trouble tickets may indicate an unmitigated or recurring problem or trend within a system that may require remediation.

2.4.5 Logs
Logs are one of the most valuable but underutilized tools to monitor controls and detect risk. Reasons for this include the fact that logs:
• Have too much data
• Can be hard to search for relevant information
• May not be enabled
• May not contain the right or appropriate data
• Are modified or deleted before being read

A log should contain a record of all important events that occur on a system, such as:
• Changes to permissions
• System startup or shutdown
• Login or logout
• Changes to data
• Errors or violations
• Job failures

A review of logs can identify risk-relevant events and can detect compliance violations, suspicious behavior, errors, probes or scans, and abnormal activity. Whether logs are reviewed can be a risk; a failure to review the logs can result in the organization not being aware of an ongoing attack. Logs must also be preserved in case they are needed at a later time for forensic analysis. Reviewing of logs may be facilitated through the use of analysis tools that can filter pertinent log data.

2.4.6 Media Reports
The media (e.g., newspapers, television, etc.) are one source of information about threats or incidents; however, a degree of professional skepticism should be exercised before acting solely based on a media account. The risk practitioner should alert those with authority to respond when a media communication could impact the organization or its employees, customers or business partners. The risk practitioner may provide advice on how to respond to a threat mentioned in the media that could affect the organization or a product or service used by the company's internal or external stakeholders. All stakeholders need to be educated on potential risk associated to the news feed by the appropriate department in the organization.

2.4.7 Observation
Watching a process as an independent observer may highlight issues that are unable to be seen as clearly when in the middle of daily operations. Observation may also identify a situation where the documented process is not being followed.

The risk practitioner should validate and, if necessary, update existing workflows during the exercise. If a workflow currently does not exist, the risk practitioner should create one during the exercise. Before the activity is completed, the risk practitioner should work with the subject matter expert and the system owner to verify the workflow is accurate.

2.4.8 Self-assessments

A primary objective of many organizations is to encourage direct management involvement in monitoring risk and control effectiveness within their areas of responsibility. A local manager has direct insight into the behavior of their staff and into daily operations. Therefore, they are well-suited to evaluate compliance with procedures, recurring problems, risk trends and vulnerabilities. The promotion of self-assessment can provide the risk team and senior management with regular reporting of risk and assurance of control effectiveness. When properly conducted, self-assessment may also reduce the need for more intense audits or system recertification.

2.4.9 Third-party Assurance

A valuable source of information about risk, control effectiveness, vulnerabilities and compliance can be obtained from third-party organizations that will perform reviews, audits, compliance verification and analysis of the organization, its processes, incidents, logs and threat environment. The organization can benefit from the expertise, objectivity and credibility that a third-party organization may provide. This can provide assurance that the organization is managing risk adequately and is compliant with good practices or standards.

2.4.10 User Feedback

System users are in the best position to know the system and its problems, vulnerabilities and shortcomings. Interacting with the users and learning the issues they face when using the system may indicate where security controls could be circumvented for convenience or improved efficiency.

2.4.11 Vendor Reports

There are many sources for information on current threats, vulnerabilities, new types of malware or new attack methods being used. Many government-sponsored computer emergency response teams (CERTs) and security vendors provide free or subscription-based reports and analysis that can be a valuable source of information for a risk practitioner.

2.4.12 Vulnerability Assessments and Penetration Tests

A vulnerability assessment is a careful, methodical review of the security controls for a system with the intent of discovering any weaknesses or potential gaps in the control framework that could allow a successful attack. A vulnerability assessment can be conducted by internal or external teams and may include test techniques such as social engineering, physical security tests, network probes and scans, and application vulnerability reviews. The vulnerability assessment helps to produce a list of potential vulnerabilities that must be reviewed, prioritized and possibly mitigated.

Possible vulnerabilities include:
- Unpatched systems
- Buffer overflows
- Susceptibility to injection attacks
- Unlocked server rooms
- Exposed cabling
- Sensitive data left on unattended desks or screens
- Open ports or services that are not required

> **A penetration test is a focused test that attempts to break into a system through a potential vulnerability.**

A penetration test can be used to validate the results of a vulnerability assessment and prove whether the controls and countermeasures used by the organization are working correctly. A penetration test identifies areas that may need immediate mitigation; where changes are required to procedures, policies, configurations, architecture; or where more staff training is required. See chapter 1, section 1.6.6 for more information on penetration tests.

2.5 RISK AND CONTROL ANALYSIS

The risk practitioner must compare the current state of risk against the desired state of risk, including a review of the effectiveness of controls to mitigate risk. The desired state of IT risk is closely linked to the risk acceptance level set by senior management. The risk practitioner must learn what the risk acceptance level of the organization is and then compare the current level of risk with the level of risk that is acceptable to management. Where the current level of risk exceeds the acceptable risk level, the risk practitioner must identify and document this finding and provide suggested means of mitigation.

2.5.1 Data Analysis

The risk practitioner has a wealth of data sources available (e.g., network devices, application logs, audit reports, etc.), but this can be a risk as much as a benefit. Too many data may hide or obscure important but less visible events, and the incorrect analysis of data may lead to an erroneous conclusion.

The challenges of data analysis start with the completeness and trustworthiness of the data:
• Are all of the data available?
• Have any of the data been altered or changed?
• Are the data in the correct format?
• Are the data based on measuring important factors?

Some of the approaches to conducting data analysis are listed in **exhibit 2.5**.

Exhibit 2.5: Methods for Uncovering Less Obvious Risk Factors	
Method	**Description**
Cause-and-effect analysis	A predictive or diagnostic analytical tool that is used to: • Explore the root causes or factors that contribute to positive or negative effects or outcomes • Identify potential risk **Note:** A typical form is the Ishikawa diagram, also known as the fishbone diagram.

Fault tree analysis	A technique that:
	• Provides a systematic description of the combination of possible occurrences in a system, which can result in an undesirable outcome (top-level event)
	• Combines hardware failures and human failures
	
	A fault tree is constructed by:
	• Relating the sequences of events that, individually or in combination, could lead to the top-level event
	• Deducing the preconditions for the:
	  – Top-level event
	  – Next levels of events, until the basic causes are identified (elements of a "perfect storm" [unlikely simultaneous occurrence of multiple events that cause an extraordinary incident])
	
	Note: The most serious outcome is selected as the top-level event.

Sensitivity analysis	A quantitative risk analysis technique that: • Helps to determine which risk factors potentially have the most impact • Examines the extent to which the uncertainty of each element affects the object under consideration when all other uncertain elements are held at their baseline values **Note:** The typical display of results is in the form of a tornado diagram.

When analyzing data to determine risk levels, the risk practitioner should be attentive to the trends of events in the data sources. A problem may be emerging that should be identified and mitigated as soon as possible. This is only possible when regular analysis is conducted so that a trend can be noticed over time.

2.5.2 Threat and Misuse Case Modeling

In chapter 1, section 1.6, the topic of understanding the mindset and approach of the adversary is discussed. The motivation and skill level of the attacker are important in determining the real existence of a threat. Threat modeling examines the nature of the threat and potential threat scenarios. This is a valuable tool in the hands of the risk practitioner and will greatly assist in risk assessment. Threat modeling is done by mapping the potential methods, approaches, steps and techniques used by an adversary to perpetrate an attack. The threat agent will often try different tools, probe for various vulnerabilities, and try several approaches—both technical and nontechnical—to compromise a system. A risk practitioner must think of all the methods and approaches a threat agent may use in attempting to penetrate a system so that adequate controls can be designed to meet possible threats. Threat modeling helps the risk practitioner and systems designers, developers and operators to build systems with attention to defensive controls, with built-in security features and as a part of a defense in depth strategy that can avoid system failures.

Threat modeling and misuse case modeling are different from use case modeling. Use case modeling examines how a system will function and provide "use" for the users. Misuse case modeling looks at all the possible errors, mistakes or ways a system can be misused. Understanding misuse cases can ensure that a system is built with resiliency and the ability to handle errors and misuse. Threat modeling examines the ways a system can be attacked and used for a purpose for which the system was never intended. An example of this is the "ping of death" attack. The Internet control message protocol (ICMP) packet was designed as a tool for system and network administrators to test network connectivity. However, by altering the size of an ICMP packet, an attacker is able to build an attack tool that could disable a victim's system. The ICMP tool was never designed for such a purpose, but the attackers learned how to use it improperly. Such threats have also used network time protocol (NTP) and domain name system (DNS) services as an attack platform.

2.5.3 Root Cause Analysis

Root cause analysis is a process of diagnosis to establish the origins of events, which can be used for learning from consequences, typically from errors and problems.

The decision of what actions to take to respond to risk are often based on the lessons learned from previous events. A prudent manager examines the root cause of an incident to discover the conditions and factors that led to the event, rather than reacts to the symptoms of the problem. Root cause analysis discovers the core, or root, of a problem and ensures that the actions taken are focused on fixing the source of the problem. For example, a review of a business process may find that the users are not compliant with the procedures and policies in place. A risk practitioner should examine why the users are not compliant before recommending enforcement of the procedure. It may be found that the procedure is outdated, flawed or unworkable and that the procedure itself should be corrected. Root cause analysis examines the reasons why a problem exists or a breach has occurred and seeks to identify and resolve any underlying issues.

One implementation of root cause analysis is a pre-mortem. A pre-mortem is a facilitated workshop where the group is told to pretend that the project has failed and then they are to discuss why it has failed. The simple but crucial cognitive impact of answering the question in this fashion instead of why it might fail, if facilitated correctly, can produce insightful, collaborative and valuable perspectives on risk.

In many cases, a risk event may be the result of several issues that act in combination with each other to result in the incident; these are called coinciding events. In this instance, the solution is more than just a single response activity or implementation of a single control.

2.5.4 Gap Analysis

The road between an organization's current risk state and its desired risk state may require several initiatives and projects to traverse. Understanding the gap between the current and desired state is to know the various projects that must be undertaken and the milestones that must be met before the goal is reached. Gap analysis is based on documenting the desired state or condition of risk that management wants to reach and then carefully analyzing and evaluating the current condition of the organization. A gap analysis identifies the gap or difference between the desired and current state so that corrective action can be taken where necessary.

The ways to measure state can be based on key goal indicators (KGIs), key performance indicators (KPIs), international standards, good practices or compliance. This process is illustrated in **exhibit 2.6**.

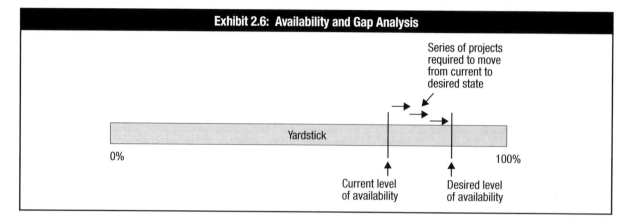

Exhibit 2.6: Availability and Gap Analysis

Series of projects required to move from current to desired state

Yardstick

0%   100%

Current level of availability

Desired level of availability

Using the gap analysis method to plan the projects and solutions needed to improve the security will ensure that projects are conducted in an orderly and logical manner and that the dependencies among the various projects are identified.

Some of the methods and artifacts used by the risk practitioner to determine the desired state of IT risk include:
• Interviews with management
• Regulations and legislation
• International standards and good practices
• Policies
• KPIs
• Key risk indicators (KRIs)

Key Performance Indicators

A KPI is a measure that determines how well the process is performing in enabling the goal to be reached. It is a good indicator of capabilities, practices and skills. It measures an activity goal, which is an action that the process owner must take to achieve effective process performance.

KPIs are useful to benchmark what the risk management goals are and whether those goals are being attained. Management sets the goals according to their risk acceptance level and desired cost-benefit analysis. The use of a KPI will indicate whether current risk levels are in accord with business objectives. For more information on KPIs, see chapter 4, section 4.2.

Key Risk Indicators

A KRI is a subset of risk indicators that are highly relevant and possess a high probability of predicting or indicating important risk.

A KRI is a good method of indicating a trend that may have the potential to result in a problem in the future. By monitoring KRIs, it may be possible to detect a pattern of activity that may result in an unacceptable level of performance in the future.

Through the comparison between current and desired state of the risk management program and the IT risk profile of the organization, the risk practitioner is able to assess the level or severity of the current risk and recommend the steps necessary to ensure that the state of risk of the organization will meet acceptable levels. For more information on KRIs, see chapter 4, section 4.1.

Key Goal Indicators

A KGI is a measure that tells management, after the fact, whether an IT process has achieved its business requirements and is usually expressed in terms of information criteria. A KGI is used to notify management on the status of critical reporting criteria.

2.6 RISK ANALYSIS METHODOLOGIES

Risk analysis is a complex and important process that is needed to provide the data necessary for the risk response activity. There are two main methods of analyzing risk: quantitative and qualitative. Each method has challenges, and many organizations use a combination or hybrid of the two methods, which is called semiquantitative.

2.6.1 Quantitative Risk Assessment

Quantitative risk assessment is based on numerical calculations, such as monetary values. It is most suitable for supporting cost-benefit analysis calculations because all IT risk can be related to a monetary value that can be compared to the cost of a control and the value of the benefit that the control would provide. The problem with quantitative risk assessment is the difficulty of placing a quantitative value on subjective elements of risk, such as reputation, morale and likelihood or impact of an event. Many events used in quantitative assessment are unpredictable, and the calculations are subjective and based more on speculation than on justifiable facts.

A quantitative risk assessment is often based on the calculation of the impact of a single risk event and on what the event would cost including direct costs, such as lost sales (including immediate and long-term impact), and indirect costs, such as loss of competitive advantage or damage to reputation.

One challenge with all risk assessment approaches is the problem of forecasting likelihood or frequency of a risk event. Using empirical or historical data, the risk practitioner may attempt to predict the average likelihood of an event over an entire population, but this is completely unreliable in an individual sense. An example is that average life expectancy of the entire population may be one figure but on an individual basis any one person may be the exception. So also in IT risk assessment, an event such as flooding or a tornado may occur at a certain average frequency, but may happen much more or less often to any one organization or location.

To properly use the data obtained from a quantitative risk assessment, the cost of a risk is often calculated on an annual basis. It is necessary to compare various events that occur at different frequencies in a similar manner using a common denominator. Because most budgets are calculated on an annual basis, calculating the cost of risk on an annual basis is also common.

Calculating the Cost (Impact/Consequence) of an Event

The calculation of the cost of an event is difficult to determine. As noted, this can be unpredictable depending on the effectiveness of the risk prevention, detection and response process; the speed at which the event was detected and contained; whether the incident gained media attention; and the fines or penalties associated with the event.

Threat/Vulnerability Pairings

Modern risk assessment understands that risk is not based on a simple threat/vulnerability pairing where a threat will attack one specific vulnerability and cause damage. Instead, many threats could attack a single vulnerability and many different vulnerabilities could be attacked by a single threat. The risk practitioner must be aware of the many relationships between threats and vulnerabilities that could be a source of risk to the organization.

2.6.2 Qualitative Risk Assessment

Qualitative risk assessment is usually based on scenarios or descriptions of situations that may have occurred or may occur. The intention of the scenarios is to elicit feedback from all the stakeholders (e.g., departments, customers, management) regarding the impact that such a scenario would have on the operations in that department. The feedback is usually not numeric and is based on a range of values such as:

- Very low
- Low
- Moderate
- High
- Very high

By communicating with all affected shareholders, the risk practitioner should be able to determine the level of risk based on feedback from all potentially affected groups.

The development of scenarios may be based on threats, vulnerabilities or asset/impact. A threat-based scenario examines a risk event from the basis of what threat sources (threat agents) exist and the threats that could be launched against the organization. The threat-based scenario would identify the potential method of attack, the vulnerabilities exploited, the intent and skill of the attacker and the potential damage to the assets affected. This method is especially beneficial when examining the emergence of new threats and determining the risk related to advanced persistent threats (APTs).

A vulnerability-based approach examines the organization's known vulnerabilities and then attempts to determine the threats that could exploit those vulnerabilities and their impact. This is especially valuable after completing a vulnerability assessment to determine the severity of the vulnerabilities.

An asset/impact approach is based on the identification of critical (availability) and sensitive (confidentiality and integrity) assets and the potential ways that an asset could be damaged. This approach is also used in BIAs. A BIA is based on determining the impact of a risk event on an asset (usually a product or service provided by the organization) over time.

Results of a Qualitative Risk Assessment

A qualitative risk assessment is typically used to create a table that compares the likelihood with the impact of a risk event. The confluence of the two factors generates the relative level of the risk. Where the level of the two factors intersect represents the relative level of the risk, so a risk that is highly likely and has a high level of impact would be identified as an area of immediate concern, whereas a risk of lower likelihood or a lower level of exposure would represent a lower level of priority for risk response and treatment.

The problem with qualitative risk assessment is that is does not provide the hard numerical values needed for cost-benefit analysis, and it also tends to overemphasize low-level risk as compared to higher levels of risk.

2.6.3 Semiquantitative Risk Assessment

Semiquantitative risk assessment combines the value of qualitative and quantitative risk assessment. A hybrid approach has the realistic input of a qualitative assessment combined with the numerical scale used to determine the impact of a quantitative risk assessment.

In a semiquantitative approach, the risk practitioner creates a range of values that are used to assess risk. For example, using a range of values from 0 to 100 creates a larger dispersion of risk assessment than a limited set of five qualitative levels and may result in a more evenly based assessment and be more readily understood by the experts who will provide input to risk assessment levels.

It is important that both qualitative and semiquantitative risk assessments support the business representatives who are providing input data with meaningful ways to determine the difference between the various risk levels. A person must understand what makes up a high versus a low level of risk—for example, financial, impact on operations, injury to personnel, etc.

When assessing impact, many criteria can be used, including:
• Financial loss
  – Fines
  – Lost sales
  – Contract penalties
• Damage to reputation
  – Brand
  – Customer confidence
• Injury or loss of life
• Lost opportunities
  – Competitive advantage
  – Research
• Lost productivity
• Cost to investigate and recover
• Future costs due to new procedures

An example of semiquantitative risk assessment is shown in **exhibit 2.7**.

Exhibit 2.7: Assessment Scale—Level of Risk			
Qualitative Values	**Semi-quantitative Values**		**Description**
Very High	96-100	10	Very high risk means that a threat event could be expected to have multiple severe or catastrophic adverse effects on organizational operations, organizational assets, individuals, other organizations, or the Nation.
High	80-95	8	High risk means that a threat event could be expected to have a severe or catastrophic adverse effect on organizational operation, organizational assets, individuals,other organizations, or the Nation.
Moderate	21-79	5	Moderate risk means that a threat event could be expected to have a serious adverse effect on organizational operations, organizational assets, individuals, or the Nation.
Low	5-20	2	Low risk means that a threat event could be expected to have a limited adverse effect on organizational operations, organizational assets, individuals, other organizations, or the Nation.
Very Low	0-4	0	Very low risk means that a threat event could be expected to have a negligible adverse effect on organizational operations, organizational assets, indivduals, other organizations, or the Nation.
Source: National Institute of Standards and Technology (NIST), *NIST Special Publication 800-30 Revision 1: Guide for Conducting Risk Assessments*, USA, 2012			

2.6.4 Risk Ranking

After the results of a risk assessment are compiled, a risk ranking is completed and will be used to direct the risk response effort. Risk ranking is derived from a combination of all the components of risk including the recognition of the threats and the characteristics and capabilities of a threat source, likelihood, vulnerabilities, severity of the vulnerability, likelihood of attack success (effectiveness of controls) and level of impact of a successful attack. When combined together, these indicate the total level of risk associated with a threat.

An example of a matrix that is used to document the risk ranking is shown in **exhibit 2.8**.

Exhibit 2.8: Template—Adversarial Risk												
1	2	3	4	5	6	7	8	9	19	11	12	13
		Threat Source Characteristics										
Threat Event	Threat Sources	Capability	Intent	Targeting	Relevance	Likelihood of Attack Initiation	Vulnerabilities and Predisposing Conditions	Severity and Pervasiveness	Likelihood Initiated Attack Succeeds	Overall Likelihood	Level of Impact	Risk
Source: NIST, *NIST Special Publication 800-30 Revision 1: Guide for Conducting Risk Assessments*, USA, 2012												

Operationally Critical Threat Asset and Vulnerability Evaluation® (OCTAVE®)

One approach to risk assessment and ranking is the use of OCTAVE. OCTAVE is used to assist an organization in understanding, assessing and addressing its information security risk from the perspective of the organization. The OCTAVE process-driven methodology is used to identify, prioritize and manage information security risk. It is intended to help organizations:[4]
- Develop qualitative risk evaluation criteria based on operational risk tolerances
- Identify assets that are critical to the mission of the organization
- Identify vulnerabilities and threats to the critical assets
- Determine and evaluate potential consequences to the organization if threats are realized
- Initiate corrective actions to mitigate risk and create practice-based protection strategy

OCTAVE focuses on critical assets and the risk to those assets. It is a comprehensive, systematic, context-driven and self-directed evaluation approach. It can help an organization to maintain a proactive security posture and apply the organizational point of view to information security risk management activities.[5]

OCTAVE:[6]
- Identifies critical information assets
- Focuses risk analysis activities on these critical assets
- Considers the relationships among critical assets, the threats to these assets and the vulnerabilities (both organizational and technological) that can expose assets to threats
- Evaluates risk in operational context, i.e., how the critical assets are used to conduct the organization's business and how they are at risk due to security threats and vulnerabilities
- Creates practice-based protection strategy for organizational improvement as well as risk mitigation plans to reduce the risk to the organization's critical assets

According to Panda,[7] the OCTAVE process is based on three primary phases:
- **Phase 1: Build asset-based threat profiles (organizational evaluation)**—The analysis team determines critical assets and what is currently being done to protect them. The security requirements for each critical asset are then identified. Finally, the organizational vulnerabilities with the existing practices and the threat profile for each critical asset are established.
- **Phase 2: Identify infrastructure vulnerabilities (technological evaluation)**—The analysis team identifies network access paths and the classes of IT components related to each critical asset. The team then determines the extent to which each class of component is resistant to network attacks and establishes the technological vulnerabilities that expose the critical assets.
- **Phase 3: Develop security strategy and mitigation plans (strategy and plan development)**—The analysis team establishes risk to the organization's critical assets based on analysis of the information gathered and decides what to do about the risk. The team creates a protection strategy for the organization and mitigation plans to address identified risks. The team also determines the "next steps" required for implementation and gains senior management's approval on the outcome of the entire process.

Measuring Risk Management Capabilities

When assessing risk, it is important to measure the capability and maturity of the risk management processes of the organization. An organization with a capable and mature risk management process is much more likely to prevent incidents, detect incidents sooner and recover rapidly from incidents.

A mature organization has defined, reliable processes that it follows consistently and continuously seeks to improve. The organization should seek to implement well-structured risk management procedures across all of its departments and regions and ensure that every business process and systems development project follows the organization's core risk management principles, policies, procedures and standards. The risk management framework should reflect the risk appetite of the organization because it is the core of the organization's risk culture.

Key elements used to measure IT risk management capability include:
- Support of senior management
- Regular communication between stakeholders
- Existence of policy, procedures and standards
- Completion of a current BIA
- Logging and monitoring of system activity
- Regular review of logs
- Scheduled risk assessments and review
- Testing of business continuity plans (BCPs) and disaster recovery plans (DRPs)
- Training of staff
- Involvement of risk principles and personnel in IT projects
- Gathering feedback from users and stakeholders
- Validating the risk appetite and risk acceptance levels
- Time to detect/resolve a security incident

Improvements in risk management capability (efficiencies and effectiveness) are based on the consistent application of the policies and procedures. If the procedures are followed on a consistent basis, then the efficiency and effectiveness of the risk management practices will improve. This will allow for review and improvement to the procedures.

Reviewing Risk Appetite Bands
Once the risk has been determined, using quantitative, qualitative or semiquantitative methods, the risk practitioner should review whether the risk levels, as seen in **exhibit 2.9**, are within the boundaries of acceptable risk. The level of acceptable risk is determined by senior management. The risk practitioner documents and reports in the risk register. Where the level of risk exceeds the boundaries set by management, the risk practitioner can provide recommendations on how to mitigate or address the risk appropriately.

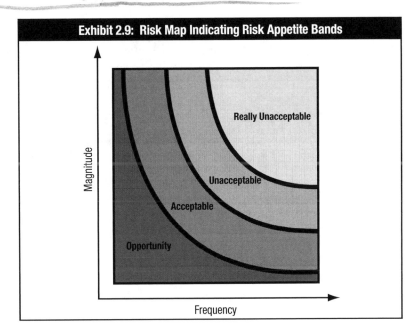

Exhibit 2.9: Risk Map Indicating Risk Appetite Bands

2.7 DOCUMENTING INCIDENT RESPONSE

When assessing risk, it is important to ensure that all IT risk has been evaluated. Some IT risk events will not apply to every organization (e.g., not all regions are subject to the same natural disasters). When an assumption is made to overlook a certain risk, it should be documented to validate that the risk event was intentionally bypassed, not just missed in the risk assessment process. Most of the risk should have been documented previously in the risk identification process, but as a deeper level of investigation and evaluation of risk is conducted, new risk may surface or those that were previously overlooked may become relevant. For this reason, the risk practitioner should reevaluate each documented risk to ensure that each risk was identified and assessed accurately based on the current risk landscape.

Risk is usually the result of a combination of several factors. Many times, a risk may be the result of preexisting or predisposing conditions that have been present for a time period but were never exploited until a later time. A risk event may also be the result of a trigger that served as the spark to ignite a risk event and cause an incident. A seemingly minor problem or vulnerability may lead to a serious problem in the event that several problems converge to result in a major outage. The risk practitioner must consider all of these scenarios when assessing risk.

The risk practitioner should pay close attention to previous incidents, audit reports and failures affecting both internal and external parties. Each event has the potential to provide a wealth of information on how to better protect systems, how to better detect an incident and how to respond more effectively. The greatest threat to the incident response process is a failure to learn from past events and to, therefore, repeat the same errors in the future. The investigation of past events may also validate the calculation of risk impact; however, it must be cautioned that no two incidents are identical and a future event may have a much more, or less, serious level of impact than a past failure.

2.8 BUSINESS-RELATED RISK

The impact of risk is a measure of the impact on the business. In the next phase of IT risk management, the risk response will have to be chosen based on business-related considerations more than on IT factors. The cost of a risk and the cost of a control is a cost to the profitability and success of the business. The impact is the damage caused to the business, not just the inconvenience to the IT department. When the IT risk assessment report is given to management at the end of this step, it must express risk in terms that management can understand and relate to and reflect items about which management is concerned. For this reason, risk should be documented in business terms and in a language that is clear and understandable to management. All documents developed during the risk assessment process should be written using common business terminology, and the risk practitioner should refrain from using terminology that is specific to IT or may be misinterpreted by management. The risk practitioner should be aware that many terms, especially acronyms that are commonly used in IT, have different meaning to other disciplines in the organization. This may result in the misunderstanding of the problem and the proposed solutions.

2.8.1 Business Processes and Initiatives

When documenting risk, it should be remembered that risk is a strategic consideration. An emphasis on current risk is incomplete without also documenting the risk related to new initiatives, technologies, lines of business and organizational changes. The IT risk assessment process must be integrated and aligned with the direction of the organization and seek to provide a secure, reliable environment for operations well into the future. IT risk management is an enterprisewide function and must, therefore, pay attention to upcoming changes and initiatives and be at the forefront of providing advice on how to manage risk even before the organization has ventured in a new direction.

Whenever a business process is being changed or revised, the risk associated with the new process should be examined. Will the new process eliminate some existing controls? Will it bypass segregation of duties? Will the ability to manage and audit risk still be available?

2.8.2 Management of IT Operations

Many elements of risk are directly a result of how an IT department is run and managed. Risk depends on the culture of the organization and the approach (attitude) of management to acknowledge and manage risk. This may be especially concerning in an IT department where an attitude of complacency and risk ignorance can result in breaches of security. It has been alleged that a majority of security incidents can, in whole or in part, be the result of the actions or inactions of IT staff. Security problems often seen within the IT department include sharing high-level accounts and passwords and the lack of control over changes to production systems. The pressure to complete the project on time or meet demands from the business can mean that shortcuts are taken that pose a significant risk to the business.

IT management must play an active role in mitigating risk and supporting risk management activities. The example set by IT management is crucial to providing a good risk culture and encouraging the integration of risk management principles into IT systems and projects.

2.9 RISK ASSOCIATED WITH ENTERPRISE ARCHITECTURE

Enterprise architecture (EA) focuses on producing a view of the current state of IT, establishing a vision for a future state and generating a strategy to get there (preferably by optimizing resources and risk while realizing benefits). This view of IT should demonstrate links between IT and organizational objectives and produce a view of current risk and controls. However, significant effort, resources and time are required to develop an EA, and generally these programs are at a low maturity state.

EAs also benefit from being informed by an understanding of organizational strategy and the views of the senior executive, which change rapidly in the current business environment and, therefore, need to be regularly reviewed. In any knowledge-based organization or an organization dependent on information systems, EA is a key component of the control environment. The EA should answer, or enable the answering of, the four AREs, as shown in **exhibit 2.10**.

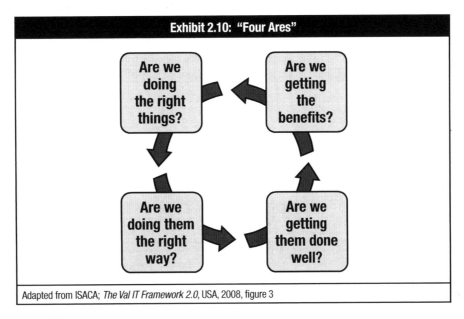

Exhibit 2.10: "Four Ares"

- Are we doing the right things?
- Are we getting the benefits?
- Are we doing them the right way?
- Are we getting them done well?

Adapted from ISACA; *The Val IT Framework 2.0*, USA, 2008, figure 3

The risk associated with an IT system is a combination of the risk associated with each element that makes up an IT system. A risk in any one location can affect the security of all other areas, and a vulnerability on one system can affect the security of all other systems. Risk management requires a complete and thorough assessment of the risk to each element of a system.

2.9.1 Hardware

IT systems and business processes rely on hardware—the equipment and devices that process, store and transmit data. Hardware includes many devices, such as:
- Central processing units (CPUs)
- Motherboards
- Random access memory (RAM)
- Read-only memory (ROM)
- Networking components (switches, routers, etc.)
- Firewalls and gateways
- Keyboards
- Monitors

Risk associated with hardware includes:
- Outdated hardware
- Poorly maintained hardware
- Misconfigured hardware
- Poor architecture
- Lack of documentation

- Lost, misplaced or stolen hardware
- Hardware that is not discarded in a secure manner
- Sniffing or capturing traffic
- Physical access
- Hardware failure
- Unauthorized hardware

2.9.2 Software

Software includes the applications, operating systems, utilities, drivers, middleware, application program interfaces (APIs), database management systems (DBMS) and network operating systems that manage data, interface between systems, provide a user interface to hardware and process transactions on behalf of the user.

Risk associated with software includes:
- Logic flaws or semantic errors
- Bugs (semantic errors)
- Lack of patching
- Lack of access control
- Disclosure of sensitive information
- Improper modification of information
- Loss of source code
- Lack of version control
- Lack of input and output validation

Operating Systems

The operating system (OS) is the core software that allows the user to interface with hardware and manages all system operations. The OS manages access to system resources and controls the behavior of system components.

Risk associated with the OS includes:
- Unpatched vulnerabilities
- Poorly written code (buffer overflows, etc.)
- Complexity
- Misconfiguration
- Weak access controls
- Lack of interoperability
- Uncontrolled changes

Applications

Applications are the face of the information system and are the mechanism by which most users can access information, perform transactions and use system features.

Risk associated with applications includes:
- Poor or no data validation
- Exposure of sensitive data (i.e., lack of encryption/obfuscation)
- Improper modification of data
- Logic flaws (i.e., logic errors)
- Software bugs (i.e., coding errors)
- Lack of logs
- Lack of version control
- Loss of source code
- Weak or lack of access control
- Lack of operability with other software
- Back doors
- Poor coding practices

2.9.3 Utilities

Utilities can include two separate areas of concern: environmental controls and those that support the use of system resources. Environmental controls include power and heating, ventilation and air conditioning systems (HVAC) that support IT operations. Utilities can also be used to refer to the utilities and drivers that support the use of system resources such as printers, keyboards, applications and networks.

Environmental Utilities

Risk associated with environmental utilities includes:
- Power interruptions
  - Loss of power
    · Surge
    · Spikes
    · Sags
    · Brownouts
    · Faults
  - Generators
    · Insufficient capacity
    · Poor maintenance
  - Batteries
    · Poorly maintained
    · Outdated
- HVAC
  - Overheating
  - Humidity problems
    · Corrosion and condensation (high humidity)
    · Static (Low humidity)
  - Clogged filters
  - Lack of maintenance
- Water
  - Loss of water (needed for cooling systems)
  - Health and safety issues
- Secure operational areas
  - Restricted access to server rooms and wiring closets
  - Secure access to power supplies, generators, elevator shafts

Software Utilities

Risk associated with utilities and drivers includes:
- Use of outdated drivers
- Unavailability of drivers
- Unpatched drivers
- Use of insecure components (older encryption algorithms)
- Unpatched vulnerabilities

2.9.4 Platforms

IT platforms vary by organization. Less complex companies may have all of their technology residing on a single platform such as the mainframe, while other organizations utilize a variety of platforms to capture, store and process information. An organization's infrastructure may be located in a single centralized location or may be decentralized in multiple locations. It may be managed by its own staff or outsourced to a service provider located within the organization or at the provider's facility.

Many IT platforms currently in use range from centralized to decentralized and from insourced to outsourced. Some organizations use a multitiered platform, such as a three-tiered infrastructure that uses middleware, to interface between legacy systems and a user-friendly front-end application. Many systems integrate with databases that may be located on different networks or use virtual environments or thin client infrastructures that reduce functionality and vulnerabilities at the client machine level.

An example of centralized architecture, shown in **exhibit 2.11**, is the use of a mainframe in which each user connects to a common processer—in this case a mainframe. The users may connect through a dumb terminal or a personal computer. In a traditional mainframe, all of the processing was done on the mainframe and the user's computer was only used for data entry and display. Currently, some pre-processing is done on the user's computer, including input validation and generating the graphical display to make the data more presentable to the user.

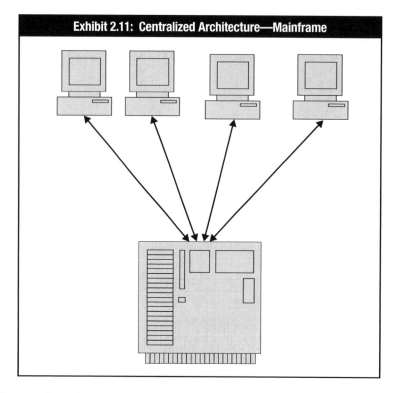

Exhibit 2.11: Centralized Architecture—Mainframe

A three-tiered architecture, shown in **exhibit 2.12**, may use middleware to provide an interface between numerous underlying systems that may be located on various servers and the user. This allows the user to have one interface with several underlying servers. This provides the user with a much easier method of accessing data from disparate systems without having to log on to each system individually.

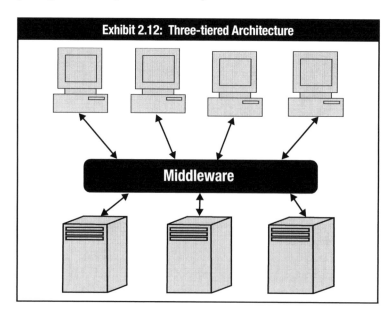

Exhibit 2.12: Three-tiered Architecture

The choice of platform or platforms that are in use is not as important as the way in which the platform(s) is managed. Some platforms are more secure than others, such as a mainframe, while others are more vulnerable, such as a web application or a kiosk operating in a public area. In some cases, a system is deployed into a vulnerable environment or a legacy system is still in use that does not have sufficient native controls. In these cases, the risk practitioner should look for compensating controls that can be used to provide additional protection for the systems.

The risk practitioner must assess the risk associated with the attitude and diligence displayed by the architects and IT operations staff to harden systems, follow good practices in change control while performing administrative functions, and ensure that the systems are tested on a scheduled basis to ensure compliance with standards, procedures and good security practices.

2.9.5 Network Components
A network is made up of many devices including cabling, repeaters, switches, routers, firewalls, gateways and wireless access points. Currently, most networks are digital, where the communications are sent via digital signals such as bits or pulses of light. Digital signals tend to be much higher quality than analog technologies, but they suffer from more attenuation than analog traffic. This means that the signal quality rapidly diminishes and may require repeaters to regenerate the signal and extend its range.

Networks are an important part of IT risk assessment because networks are often the target of an attack or the channel used in the attack on a system or application. The network itself is often essential to business operations, and many business processes rely on the availability of the network. The simple definition of a network is a system of interconnected computers and the communication equipment used to connect them. Networks may range in size from a small personal area network of a headset talking to a smartphone over Bluetooth®, to a global area network that links devices together worldwide, such as the Internet.

Networks are used for many purposes, including:
• Transferring data between individuals
• Transferring data between applications
• Controlling and monitoring of remote equipment (supervisory control and data acquisition [SCADA] networks)
   Backing up data (storage area networks [SANS])
• Enabling communication between devices (Bluetooth)

When assessing network-based risk, the risk practitioner must examine the risk associated with:
• Network configuration and management, including the recognition of the criticality of network operations
• Network equipment protection
• The use of layered defense (defense in depth)
• Suitable levels of redundancy
• Availability of bandwidth
• Use of encryption for transmission of sensitive data
• Encryption key management   administration of key
• Use of certificates to support public key infrastructure (PKI)
• Damage to cabling and network equipment
• Tapping network connections and eavesdropping on communications
• Choice of network architecture
• Documentation of network architecture

Cabling
Networks may be connected using cable or wireless technologies. Cabling comes in several forms including unshielded twisted pair (UTP), coaxial or fiber. Each type of cable has its own benefits and vulnerabilities, and the risk practitioner should ensure that the appropriate cabling is being used for the communications required.

The least expensive option used for many local area networks (LANs) is UTP cable with a grade of category 5e (CAT5e) or category 6 (CAT6). Category (CAT7) is available, but not yet in common use. CAT7 cable is a shielded cable that protects each pair of wires and the cable itself, thereby reducing noise and cross talk for ultra-high speed Ethernet. CAT5e and CAT6 cables are relatively easy-to-use and inexpensive cabling options that support fast Ethernet.

The concerns for the risk practitioner regarding cabling include:
• Physical security of cabling
• Cable exceeding the approved length of the cable runs (100 meters for CAT5e, 55 meters for CAT6)
• Protection from damage to cabling (conduit)
• Use in an area of high radio frequency interference (RFI) (may require shielding)
• Use of cable that is not of suitable standard (e.g., CAT3)
• Ensuring use of plenum-rated cable where required
• Improper terminations of cable on connectors
• Lack of cabling records

Repeaters
Repeaters are used to extend the length of a signal being transmitted over cable or wireless networks. As the signal traverses the network, it suffers from attenuation, and the repeater will be placed at a location to receive the input signal and regenerate the signal and send it on further. The advantage of a repeater is that it can filter out some noise or errors that may be affecting traffic. The distance between repeaters is based on the type of cabling or technology being used and the operational environment. For example, as the quality of fiber has increased over the years, it has meant that the distance between repeaters has grown substantially. In an environment with many concrete walls or other interference or where a higher frequency wireless channel is being used, repeaters may be required more often.

A risk associated with repeaters is ensuring that enough repeaters are in use to provide a clean, error-free signal. Another risk is a wireless repeater providing a strong signal into areas outside the perimeter of the organization's facilities that could allow unauthorized access.

Switches
Several terms used in network communications are quite broad, such as switches, firewalls and virtual private networks (VPNs). There are many types of switches with various functions, just as there are several generations of firewalls—many of which are considerably different in their function and operation.

Switches are used to connect devices together, and current switches perform the functions formerly provided by hubs and bridges (layer two). Switches forward packets to a destination, can perform routing functions (a layer three switch), address translation and balancing (layer four), and perform load balancing (a layer seven switch).

Switches can be used to connect networks but also segment and divide networks through configurations such as a virtual LAN (VLAN), where different devices physically connected onto one switch may be set up to act as completely separate networks.

Switches can be wired or wireless, small and inexpensive, or large and very expensive. Each organization must determine what equipment is suitable to meet its network requirements.

The risk associated with switches includes:
• Physical protection of the switch
• Ensuring the proper configuration of the switch
• Documentation
• Being a single point of failure

Routers
The purpose of a router is to connect multiple networks together and forward incoming packets in the direction of the destination IP address that is in the packet header. A router can vary in size, from a wireless router in a home that connects computers to the Internet, to large core routers that are responsible for the routing of high-speed Internet communications. The router builds a routing table using routing protocols, such as Border Gateway Protocol (the first routers were called gateways), and learns how best to direct traffic toward its destination. In many ways, a router is like a traffic circle used to handle vehicular traffic. Traffic enters the circle and then routes off on the road that is best toward the vehicle's ultimate destination. The intelligence of the routing decision is similar to the decision made by a driver—to take the shortest route, the fastest route, the route with least construction delays, etc.; therefore, a router may direct traffic over different routes depending on traffic volumes and congestion.

Just like on the road, some traffic has labels that indicate it has a priority, such as emergency vehicles that are permitted to bypass other traffic. The label, like the siren on an emergency vehicle, indicates that this packet should be processed quickly and not be held back by other packets waiting to be processed. The delay in processing is called latency, and this would severely affect the quality of some types of traffic such as voice over IP (VoIP). The labeling of the traffic allows for provision of quality of service (QoS) and class of service (CoS) traffic management. QoS usually is used to provide a guaranteed bandwidth for traffic volumes, and CoS is used to grant priority to certain packets over others (e.g., voice packets over regular data packets). This also prevents the jitter—the variation in the arrival time of a packet due to different routing, latency, etc.—that would affect data traffic from impacting voice communications.

The risk associated with a router includes:
• Improper configuration
• Use of weak protocols (RIPv1)
• Software bugs
• Unpatched systems
• Physical security

Firewalls

A firewall is defined as a system or combination of systems that enforces a boundary between two or more networks, typically forming a barrier between a secure and an open environment, such as the Internet. The risk practitioner must remember that firewalls should be used to control both incoming and outgoing traffic. The term "firewall" refers to many types of technology that operate at various layers of the open systems interconnection (OSI) model and have evolved significantly over the past years. Types of firewalls are described in **exhibit 2.13**.

Exhibit 2.13: Firewall Types	
First generation	A simple packet-filtering router that examines individual packets and enforces rules based on addresses, protocols and ports.
Second generation	Keeps track of all connections in a state table. This allows it to enforce rules based on packets in the context of the communications session.
Third generation	Operates at layer seven (the application layer) and is able to examine the actual protocol being used for communications, such as Hypertext Transfer Protocol (HTTP). These firewalls are much more sensitive to suspicious activity related to the content of the message itself, not just the address information.
Next generation	Sometimes called deep packet inspection—is an enhancement to third generation firewalls and brings in the functionality of an intrusion prevention system (IPS) and will often inspect Secure Sockets Layer (SSL) or Secure Shell (SSH) connections.

It is important to ensure that the firewall configurations are backed up on a regular basis, reviewed to ensure that all rules are in the correct order and are documented, and tested on a scheduled basis. Any changes to the firewall configuration or rules should be subject to the change management process of the organization.

A firewall is only as good as the person that manages it. An important role of the risk practitioner is to verify that the staff members managing the firewalls and other network devices are knowledgeable, trained and supervised. Firewall logs must be reviewed on a regular basis to detect any suspicious activity.

Proxy

A proxy is a device that acts as an intermediary between two communicating parties. The proxy acts to the party on each side of the communication as if it were the other end host. This allows it to filter and examine suspicious activity, protect internal resources and take action if unacceptable activity is occurring. A gateway is a type of proxy that controls traffic through a gate or security perimeter.

Domain Name System

For most people, the domain name system (DNS) is the mechanism that makes the Internet work. Without the DNS, they cannot navigate to their destination web site without knowing the Internet protocol (IP) address of the web site they wanted to visit. The DNS provides a simple cross-reference that is used to associate a normal name with the

IP address used by network devices. The IP address is a logical address given to a web site based on the addresses allocated to the Internet service provider (ISP). An individual can purchase a name for his/her web site that is more in line with his/her name, the name of the organization or the marketing program. For example, the IP address for ISACA.org is 12.239.13.0. A person wanting to visit the ISACA web site could simply type in http://12.239.13.0/. However, it is much easier to remember and enter *www.isaca.org*.

DNS is a tree structure that resolves the addresses so that a network device that receives a request from a person entering *www.isaca.org* will be able to look up the IPv4 address and then route the request properly through the network. When a network device does not know the IP address associated with a web name, it sends a DNS request up through the DNS tree to a higher level DNS resolver. The reply is then sent to the requesting device using User Datagram Protocol (UDP) over port 53. The requesting device will then store the IP address and name for future use.

There have been many attacks on the Internet using DNS over the years. People have sent false DNS replies to misroute traffic. DNS replies have been used in amplification attacks to flood a victim's system. Cybersquatting occurs when individuals reserve the names of sites that another organization may want and will only agree to sell that name to the organization at a higher price. DNS can also be used to learn information about a company and its administrators that can be used in attacks.

Wireless Access Points
Wireless devices offer a level of flexibility and ease of use not possible with traditional cable-based networks. With a wireless access network, users can move around the office and log in from multiple locations without requiring a network cable and port. The risk associated with wireless networks is primarily based on the threat of unauthorized people being able to log in. This requires segmenting the wireless access into a separate network and implementing a strong password requirement for network access. Placement of the wireless access points in a location that is not subject to interference from other devices or near a window that would broadcast a strong signal outside of the organization's facilities is also important. The installation of rogue or unauthorized wireless access points on the network is also a serious threat because a rogue device may permit a person to access the network directly without having to connect through a secure device or to bypass firewalls or other layers of defense.

Other Network Devices
Besides the network devices listed above, there are many other types of networking equipment and protocols used for network communications. The proper configuration of these devices is a critical part of secure communications and network reliability and stability. The choice of communication protocols used is important because some protocols are more secure than others. A common problem related to network security is unpatched systems or protocols, such as the network time protocol (NTP)-based attacks of late 2013 that resulted from many organizations having unpatched NTP implementations.

2.9.6 Network Architecture
Many network architectures in use today range from older bus networks to high-speed fiber backbones. The choice of network architecture is based on the number of devices (e.g., an old peer-to-peer network could only handle 8 to 12 devices), distance (e.g., LAN versus wide area networks [WANs]) and communications requirements of the network (e.g., speed, confidentiality, reliability).

Types of Network Topologies
Exhibit 2.14 illustrates commonly used physical network topologies.

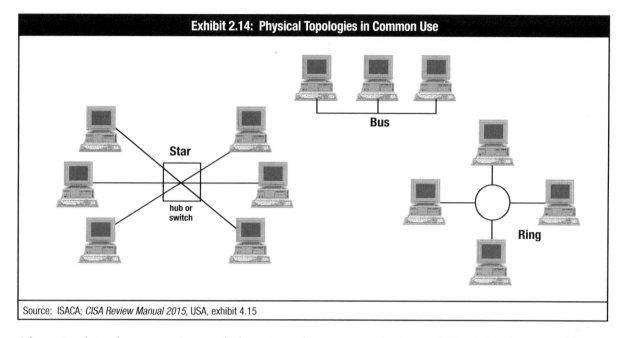

Source: ISACA; *CISA Review Manual 2015*, USA, exhibit 4.15

A bus network topology connects every device onto one bus or communications path. The risk is that a cut cable may result in total network failure, and it is relatively easy to sniff a bus network. This is one reason why an encrypted VPN must be used on cable modem Internet access.

In a star network topology, every device is connected to a central switch. This is more efficient than a bus and somewhat more difficult to eavesdrop on or sniff communications. However, the central switch is a single point of failure.

A tree network topology is a series of star networks arranged with branches to other star networks in a tree-type structure. A cut link between the branches of the tree could cause isolation of that branch, but this type of network is very scalable.

A ring network topology is used in backbones and areas where reliable high-speed communications and fault tolerance is desired. A ring connects every device into one ring and passes traffic from device to device around the ring. A ring is excellent for busy networks where a switched Ethernet would be inefficient and lead to too many collisions or traffic congestion. A ring is more expensive that most other technologies, however.

A mesh network topology is used where high availability is required. Many devices are connected to many other devices in a mesh, so that traffic can route around a failure in any part of the network. The Internet itself is a partial mesh and was built to survive massive failures of any one part of the network and still allow communications over the rest of the Internet. The risk with a mesh network is cost and the difficulty of adding new components to the network (scalability).

Any network can fail, so the risk practitioner must consider the impact of any failure and the ways that a failure can be mitigated.

Local Area Network
A LAN is defined as a communication network that serves several users within a specified geographic area, such as a building or a department. Shared data are stored in a file server that acts as a remote disk drive for all users in the network. A LAN is often architected as a star or tree, but to connect LANs together, a WAN is used.

Wide Area Network
A WAN is defined as a computer network connecting different remote locations that may range from short distances, such as a floor or building, to extremely long transmissions that encompass a large region or several countries.

There are several different topologies that can be used when a WAN must connect networks over a large geographic area, such as leased lines, packet-switching networks, satellite and wireless.

Leased Lines
The original method of connecting devices together is through a leased line. A leased line is leased or rented from a supplier (telecommunications carrier) and is provided for the sole use of the organization that leases the line. It is, therefore, a private network. These are available in a variety of speeds and configurations, ranging from a 56 kilobytes/second line to integrated services for digital network (ISDN) and fiber. Because this is a dedicated communications facility, they are expensive and also subject to being a single point of failure, but the telecommunications provider often guarantees its level of service.

Packet-switching Networks
The development of packet-switching networks allows a communications network to be shared by multiple organizations, and thereby, reduces the cost considerably. Packet-switching networks are a type of partial mesh so that they are fairly resistant to total network failure. Some of the technologies used are X.25, frame relay and asynchronous transfer mode (ATM). Some of the benefits available with these networks include QoS and CoS features that guarantee a certain amount of bandwidth availability to the organization and give priority to certain types of traffic. Through the implementation of multiprotocol label switching (MPLS), it is also possible to engineer the route that traffic takes through the network.

A problem with packet-switching networks is latency and jitter. Some packets can be received out of order. This could impact some streaming communications and affect traffic quality. A solution is to implement permanent virtual circuits (PVCs) that mandate the route that traffic takes through the network. This prevents jitter and sets a standard level of communications quality.

The single point of failure on a packet-switching network is the link between the organization and point at which the organization connects to the telecommunications provider.

Microwave
To communicate between different locations, an organization may use a microwave link. Microwave is a line-of-sight technology where the sending and receiving stations need a clear line of sight between each other. A building, tree, strong wind or other obstacle can affect the reliability of microwave communications. Microwave is also licensed in many countries, and the organization must purchase the right to broadcast on the desired frequency. Microwave also has the risk of being intercepted or monitored by an adversary.

Optical
Another option similar to microwave but based on laser technology is optical communications. The sending and receiving locations have a laser transceiver that can transmit fairly high-speed communications. Optical is also line of sight, but it is not licensed in most countries.

Satellite
Satellite communications has enabled communications from remote areas where it was not previously possible to provide other forms of communications. Satellite is a line-of-sight technology and is affected by wind, heavy rain or snow, dust, or solar flares. Although satellite signals can carry large amounts of information at a time, the disadvantage is a bigger delay due to the "jump" from the earth to the satellite and back (estimated at about 300 milliseconds).

Virtual Private Network
A VPN is designed to be a secure private network that uses the public telecommunications infrastructure to transmit data. The Internet is an insecure place, and it is fairly easy to listen in on traffic passing over the Internet or other connections such as wireless, satellite or microwave. The use of a VPN to create a tunnel for the exclusive use of the two end points is a wise decision. Keep in mind, however, that not all tunnels are encrypted. Layer 2 forwarding (L2F), layer 2 tunneling protocol (L2TP) and point-to-point tunneling protocol (PPTP) do not provide encryption. Using L2TP for remote administration (e.g., simple network management protocol [SMNP]) or other such services requires that the L2TP traffic be routed over IP security for confidentiality.

The end points of a VPN tunnel should be located in a place that will allow the traffic passing over the VPN to be examined after it has been decrypted. Otherwise, malicious traffic traversing the VPN could circumvent firewalls or other network defenses.

Length and Location

Regardless of the network architecture in use, the risk practitioner should review the risk associated with the implementation, especially regarding the suitability of the network architecture in use; the proper configuration and management of the network devices; and the ongoing monitoring of network performance, vulnerability assessments and incident management. Networks that exceed standard operating distances should be rearchitected, and networks that operate in hostile environments with high levels of interference should be reengineered to use a better technology.

Encryption

Encryption is a vital part of secure communications and is used widely in many products and systems throughout an organization ranging from email to VPNs, access control, backup tapes and hard disk storage. The secure use of encryption requires the correct choice of algorithm, the protection of the encryption keys, the randomness of key generation and limits on the length of time a key is used before being changed. In many cases, the breach of an implementation of encryption is due to user misuse or an error in the implementation, not to a problem with the algorithm. The risk practitioner is concerned with how keys are generated and stored, the training of the users, and ensuring that the algorithms adhere to current standards.

Demilitarized Zone

The area of the network that is accessible to outsiders through the Internet is isolated into a demilitarized zone (DMZ). This prevents an attacker from having direct access into internal systems. All devices in the DMZ are hardened with all unnecessary functionality disabled. Such devices are often referred to as a bastion host. Intrusion detection systems (IDSs) and intrusion prevention systems (IPSs) monitor, record and may block suspicious activity. The application firewall in the DMZ is behind a packet filtering router to clear out most of the bad traffic before it gets to the application firewall.

As can be seen in **exhibit 2.15**, a user from the Internet can only access the organization through the firewall. The firewall ensures that traffic from the outside is routed into the DMZ where the web server is located. Nothing valuable is kept in a DMZ because it is subject to attack and compromise from the outside. Outside users cannot access the internal network directly, thus providing the internal network with some measure of protection.

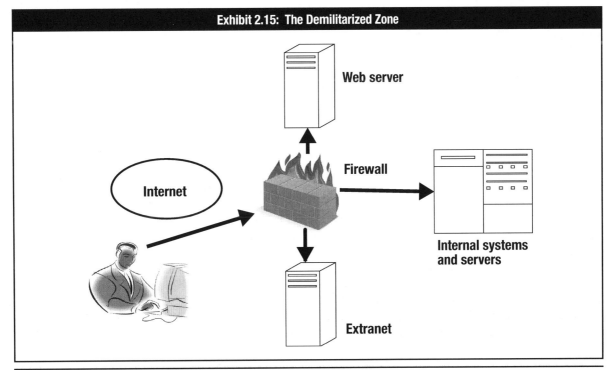

Exhibit 2.15: The Demilitarized Zone

Extranet

An extranet is a network that is, like the DMZ, accessible from outside of the organization. Extranet is used for trusted communications, such as communicating with business partners. People accessing the DMZ will often have to provide two–factor authentication. The greatest risk associated with a DMZ or extranet is misconfiguration of the devices located in those zones.

User Interfaces

The way a user interfaces with an application or a system is an important part of security and the risk profile. A user interface (UI) should restrict a user from accessing functions, data or services that are not allowed. This can be done through the use of a constrained UI such as a keyboard with limited options to choose from (such as on an automatic teller machine [ATM]) or using a drop-down menu that lists only the options a user can select. Another way to restrict a user from seeing sensitive data is to mask or encrypt the data being displayed to the user.

2.10 DATA MANAGEMENT

The risk practitioner should review and assess the data ownership and management process of the organization, including the protection of data from improper disclosure, modification or deletion. A lack of clear ownership of the data may be a significant risk. The rules for access should be maintained and enforced, and all staff should be aware of the risk associated with improper data management (such as data leakage) and their requirement to comply with the policies and standards of the organization. The risk practitioner must pay considerable attention when assessing the risk associated with data management to ensure compliance with data management policies and procedures.

2.11 NEW THREATS AND VULNERABILITIES

The risk environment changes as organizations and technologies change and as threat actors focus on new areas to attack. Systems and applications that were once thought to be secure are often found to be vulnerable at a later time. This requires the ongoing monitoring and evaluation of risk that is discussed in chapter 4, Risk and Control Monitoring and Reporting. While performing the IT risk assessment, the risk practitioner must ensure that new and emerging risk is identified and evaluated and that the organization is aware of and watching for emerging threats and vulnerabilities. This includes monitoring vendor alerts, reports from CERTs and media stories. When a new issue appears in the media, the risk practitioner should work with the business and system owner to perform a threat analysis and determine if and how the organization should respond. Responses include developing a mitigation plan or providing the necessary details to management so they can develop a response to the appropriate audience. Not being watchful of new issues or being unprepared to respond to them after they occur is a significant risk.

2.12 EMERGING TECHNOLOGIES

New technologies are coming on the market continuously. The pressure to implement new technology is often influenced by inflated expectations of its utility and maturity and a focus on product functionality without attention to addressing product security requirements. The desire to deploy new unproven technologies often causes organizations to lose sight of the business risk involved. The control for pressure to implement new technologies is to plan ahead. Foresight and forecasting practices and techniques are risk management activities that have a longer-term outlook. Emerging technologies often provide indicators years in advance of their potential. It is the job of the risk practitioner to consider the potential risk and controls for the application of these technologies that may present value to the organization. The risk practitioner should evaluate and assess the approach of the organization to accepting new technologies and the attitude of the security team and IT operations teams toward reviewing and securing the new technologies as they become available. It is not good enough to wait until a technology has been implemented before actually reviewing and securing it. A well-managed change control process ensures that new technologies are not implemented until the security team has been able to validate the security impact of the change and enable appropriate controls. A serious mistake is to ignore the new technologies and to not realize that the new technologies are possibly already in use. In most cases, if the IT department is unwilling to review and integrate new technologies, the business will find ways to implement them regardless, putting the organization in a vulnerable position.

2.13 INDUSTRY TRENDS

Marketing staff are concerned with changes in trends, and a failure to see a changing trend can result in an organization losing a substantial percentage of their market share within a very short time period. This also applies to IT. A failure by the IT department to adapt to or support a new business model may result in substantial loss to the organization. For example, when some telecommunications providers switched to per second instead of per minute billing, it forced all of the other telecommunications providers to change their systems quickly or lose a large percentage of their customers.

A risk practitioner should assess the maturity of the IT department and the organization as a whole toward monitoring and adapting to new market trends. A lack of flexibility or poor communication between the business units and IT could be a risk factor.

2.14 THIRD-PARTY MANAGEMENT

Organizations are being courted to outsource business functions, including IT and security, to outside parties that promise to address the requirements of the organization in a cheaper and more efficient manner. For some organizations, this is an ideal solution that can allow them to focus on their core business and not have to support other peripheral functions. The use of outsourcing for applications development, security monitoring and cloud-based services is an attractive option, but it also comes with risk. If an organization outsources data storage, for example, the ownership and liability for a data breach remains with the organization that utilizes the outsourcing service. Any breach is still the responsibility of the outsourcing company even though the data breach was not under its direct control. This also applies to applications development. An application supplier will often provide an executable but not the source code necessary to maintain or update the application. If the supplier subsequently goes out of business or fails to support the product, the organization that purchased the product may not be able to maintain or update the application as needed. The purchasing organization may insist that the supplier keeps a copy of the source code with a trusted third party so that the purchaser can obtain a copy in the event of a failure by the supplier to comply with contractual requirements.

If a third party is going to perform work for the organization, nondisclosure agreements (NDAs) are necessary to protect the intellectual property of the organization from being disclosed to unauthorized personnel. This is especially relevant when a penetration test or other assessment has been done that could be used to perpetrate an attack on the organization.

2.14.1 Outsourcing

The organization that uses a service provider still remains the owner and liable for the protection of their information. Any decision to outsource data management or technology functions should be carefully considered from a risk and regulatory perspective, and the outsourcing contracts should explicitly address all security requirements, including provisions for reporting and liability.

Outsourcing services, such as the cloud, can fail like any other service, so backups and DRPs must be in place. The outsourcing organization should have a method of ensuring that the provider is compliant with good practices in securing their data, such as through external audits (e.g., Statement on Standards for Attestation Engagements Number 16 [SSAE 16] reports) or having the right to audit.

Legal requirements must also address the jurisdiction for any complaints or disagreements with the provider, such as which court would have jurisdiction to hear a complaint. The agreements must also address the regulations in the host country regarding disclosure to law enforcement, and transborder data transmission and storage.

While the term "cloud" is relatively new, it is a service that has been around for many years and was formerly known as remote data hosting (such as on a mainframe). The cloud refers to the use of the Internet (usually drawn as a cloud in network diagrams) to transmit data to a remote location for storage or processing. There are many excellent cloud services available such as software as a service (SaaS), storage on demand, infrastructure as a service (IaaS) and platform as a service (PaaS). The issues are the same from a risk perspective as for any other outsourcing initiative.

The risk practitioner should ensure that security and regulatory requirements are addressed in all agreements with suppliers and service providers.

2.14.2 Contractual Requirements

When a contract is in place between two organizations, the organizations have the responsibility to comply with the terms of the contract. These can include right to audit clauses; security and BCP/DRP reviews; staffing reviews; regulatory review; outsourcers and third-party affiliate review; right for early termination if provider is acquired or acquires a company that does not meet regulatory, security and continuity requirements; service level agreements (SLAs) for activities including, but not limited to, incident response time frames, reporting requirements and failure to comply penalties; and maintenance agreements. The risk practitioner should ensure that there is regular communication between the supplier and the liaison within the outsourcing organization to identify and resolve any issues. Any areas of noncompliance should be reported to management.

2.14.3 Service Level Agreements and Contractual Requirements With Customers

An organization that does not meet its contractual requirements with customers is in danger of losing business or facing penalties, as well as suffering from reputational damage. The risk practitioner must assess the risk of the organization not being able to meet contractual requirements, such as through loss of key personnel, inability to respond promptly to an incident or conflicting demands from multiple customers.

2.15 PROJECT AND PROGRAM MANAGEMENT

There is much risk associated with project and program management. Program management is the management of a series of related projects toward the attainment of a business objective. Many IT projects fail for various reasons, such as:

- Unclear or changing requirements
- Scope creep (additional requirements)
- Lack of budget
- Lack of skilled resources
- Problems with technology
- Delays in delivery of supporting elements/equipment
- Unrealistic time lines (push to market)
- Lack of progress reporting

When a project is at risk, it is important to identify the root cause of the problem and take steps to address those problems as soon as possible. Lack of good project management can lead to:

- Loss of business
- Loss of competitive advantage
- Low morale among staff members
- Inefficient processes
- Lack of testing of new systems or changes to existing systems
- Impact on other business operations
- Failure to meet SLAs or contractual requirements
- Failure to comply with laws and regulations

When a project is at risk, the entire program may also be at risk because the program may rely on each project completing on time and on schedule.

2.15.1 The System Development Life Cycle

One method used to support systems development and IT projects is the system development life cycle (SDLC). The SDLC describes project management as a series of phases that provides structure and auditability to a project. An example of the SDLC is seen in **exhibit 2.16**.

Exhibit 2.16: Characteristics of the SDLC Phases		
SDLC Phase	**Phase Characteristics**	**Support from Risk Management Activities**
Phase 1—Initiation	The need for an IT system is expressed and the purpose and scope of the IT system is documented.	Identified risk is used to support the development of the system requirements, including security requirements and a security concept of operations (strategy).
Phase 2—Development or Acquisition	The IT system is designed, purchased, programmed, developed, or otherwise constructed.	Risk identified during this phase can be used to support the security analyses of the IT system that may lead to architecture and design trade-offs during system development.
Phase 3—Implementation	The system security features should be configured, enabled, tested and verified.	The risk management process supports implementation against its requirements and within its modeled operational environment. Decisions regarding risk identified must be made prior to system operation.
Phase 4—Operation or Maintenance	The system performs its functions. Typically the system will undergo periodic updates or changes to hardware and software; the system may also be altered in less obvious ways due to changes to organizational processes, policies and procedures.	Risk management activities are performed for periodic system reauthorization (or reaccreditation or whenever major changes are made to an IT system in its operational, production environment (e.g., new systems interfaces).
Phase 5—Disposal	This phase may involve the disposition of information, hardware and software. Activities may include moving, archiving, discarding or destroying information and sanitizing the hardware and software.	Risk management activities are performed for system components that will be disposed of or replaced to ensure that the hardware and software are properly disposed of, that residual data are appropriately handled, and that system migration is conducted in a secure and systematic manner.
Source: ISACA, *CISM Review Manual 2015*, USA, exhibit 2.29		

Regardless of the methodology used for project management, the core principles are the same. Proper oversight, clear requirements, user involvement, communications between team members and users, and regular review of project progress are all critical to project success. The risk practitioner will often find projects that are at risk of failure and then must identify the cause of the failure, recommend a solution and report the risk to management.

System development projects fail for many reasons, including:
• The requirements may change.
  – Scope creep
  – New business priorities
  – Poorly understood initial requirements
• Resources (trained staff or budget) may not be available.
• Dependencies (e.g., suppliers, outsourcers, communications providers, technology) may not deliver on time.
• Technology may not work as anticipated.
• Project complexity is underestimated.
• Resources are poorly managed.
  – Lack of leadership
  – Lack of accountability
  – Lack of oversight
• Symptoms of failure are not recognized.
• There is a lack of coordination with suppliers.

Regardless of the reason for the problems the project is experiencing, the risk to the organization is considerable for project failure. Project failure may result in:
• Indirect financial loss (loss of competitive advantage)
• Direct financial loss (violations of contract or SLAs)
• Inability to adjust to a changing operational environment
• Damage to reputation
• Dissatisfaction among the project team (decreased morale)

The identification of a project that is at risk of failure is important so that corrective action can be implemented as soon as possible. The challenge of identifying project risk is the problem of obtaining accurate data on project status and being able to identify the root cause of the project problem. Proper and engaged oversight and monitoring of the project and validation of progress reports may be required to prevent inaccurate project status reporting. Doing an objective risk assessment is often instrumental in determining true project status.

Lack of project activity coordination throughout the engagement can pose a potentially severe risk to the project's ability to be on time, on budget with the necessary resources and able to meet business expectations. That concern can manifest itself in skipping critical steps (such as testing) at the end of the project to get back on schedule, which can lead to production failures, inaccurate reporting and data processing, or missing functionality as the project is implemented.

The risk practitioner should validate that project design, development and testing did not solely focus on the business functions of the project but that it also validated that security requirements were built in and tested to reduce the risk associated with confidentiality, availability and integrity of the systems and the data it processes. The risk practitioner should ensure that risk management is being conducted throughout each step of the SDLC and that security requirements and risk are being integrated into the project life cycle. Early planning and awareness will result in cost and time saving through proper risk management planning. Security discussions should be performed as part of (not separate from) the development project to ensure solid understanding among project personnel of business decisions and their risk implications to the overall development project.[8]

Key tasks to perform during the SDLC to expressly document the risk associated with the development of a new program include:
• Security categorization of the proposed system—What are its availability, confidentiality and integrity requirements?
• BIA—What impact would an outage have on critical business processes?
• Privacy impact assessment—What laws or regulations apply? What sensitive data are processed, stored or transmitted by this system?
• Use of a secure information systems development process—Is there security training for all development staff and a secure environment? Are there secure code practices?
• Awareness of vulnerabilities with selected technology or operational environment

The project must also provide the ability to audit and review the applications or systems after implemented to ensure proper operation of the system.

2.16 BUSINESS CONTINUITY AND DISASTER RECOVERY MANAGEMENT

All systems will fail from time to time, but the impact of system failure can be mitigated through effective incident response, business continuity planning and disaster recovery planning.

The purpose of BCPs and DRPs is to enable a business to continue offering critical services in the event of a disruption and to survive a disastrous interruption to activities. Rigorous planning and commitment of resources is necessary to adequately plan for such an event.

The first step in preparing a new BCP is to identify the business processes of strategic importance—those key processes that are responsible for both the permanent growth of the business and for the fulfillment of the business goals. Ideally, the BCP/DRP should be supported by a formal executive policy that states the organization's overall target for recovery and empowers those people involved in developing, testing and maintaining the plans.

Based on the key processes, the risk management process should begin with a risk assessment. The risk is directly proportional to the impact on the organization and the probability of occurrence of the perceived threat. Thus, the result of the risk assessment should be the identification of the following:
• The human resources, data, infrastructure elements and other resources (including those provided by third parties) that support the key processes
• A list of potential vulnerabilities—the dangers or threats to the organization
• The estimated probability of the occurrence of these threats
• The efficiency and effectiveness of existing risk mitigation controls (risk countermeasures)

Business continuity planning is primarily the responsibility of senior management, because they are entrusted with safeguarding the assets and the viability of the organization, as defined in the BCP/DRP policy. The BCP is generally followed by the business and supporting units, to provide a reduced but sufficient level of functionality in the business operations immediately after encountering an interruption, while recovery is taking place. The plan should address all functions and assets required to continue as a viable organization. This includes continuity procedures determined necessary to survive and minimize the consequences of business interruption.

Business continuity planning takes into consideration:
• Those critical operations that are necessary to the survival of the organization
• The human/material resources supporting them

Besides the plan for the continuity of operations, the BCP includes:
• The DRP that is used to recover a facility rendered inoperable, including relocating operations
• The restoration plan that is used to return operations to normality, whether in a restored or new facility

Depending on the complexity of the organization, there could be one or more plans to address the various aspects of business continuity and disaster recovery. These plans do not necessarily have to be integrated into one single plan. However, each has to be consistent with other plans to have a viable business continuity strategy.

It is highly desirable to have a single integrated plan to ensure that:
• There is proper coordination among various plan components.
• Resources committed are used in the most effective way and there is reasonable confidence that, through its application, the organization will survive a disruption.

Even if similar processes of the same organization are handled at a different geographic location, the BCP and DRP solutions may be different for different scenarios. Solutions may be different due to contractual requirements. For example, the same organization is processing an online transaction for one client and the back office is processing for another client. A BCP solution for the online service will be significantly different than one for the back office processing.

2.16.1 Incident Management

Incident management starts with the preparation and planning that build an incident response plan (IRP). The organization should prevent incidents where possible, but also have the controls in place to detect and respond to an incident when it occurs. The sooner an incident can be detected, the easier it is to contain the incident and minimize the damage or impact caused by the incident. The response team must have training on the IRP and the skills necessary to investigate the incident. Incident response teams (IRTs) generally consist of employees assigned specific roles during various types of incidents. However, external resources may be brought in to assist, especially when there is a need to perform forensic investigations.

The primary focus of incident management is to get the organization's affected systems and operations back into normal service as quickly as possible. This can impact the ability to secure evidence associated with the incident because evidence may be lost while returning to service. A detailed IRP and regular testing that involves the business will increase awareness and ensure the appropriate time is granted for evidence collection and analysis while still meeting the business recovery time objectives (RTOs)/recovery point objectives (RPOs). If the business does not understand the need to collect evidence or the IRT does not know how to collect and preserve evidence, the organization is at a greater risk to be liable for the event, and the attacker has a greater chance to avoid prosecution.

Each incident must be examined to extract the lessons learned to improve the prevention, detection and recovery from future incidents of a similar nature.

2.16.2 Business Continuity Plan

Some incidents cannot be resolved in a timely manner. A serious incident may have the potential to disrupt business operations for an unacceptable period of time. This is where business continuity management comes into action. Business continuity focuses on continuing critical business operations in the event of a crisis and having plans in place to support those operations until the business can return to normal operations. The recovery of critical business processes may be through an alternate process, including:
• A manual process or outsourced support
• Having sufficient inventory on hand to support operations
• Using facilities available at another office or location
• Displacing less critical work with more critical functions

The core source of data used in business continuity planning is a BIA. A BIA examines the services and operations of the organization to identify the critical time lines for those services and products. A BIA examines the impact of an outage on the business over the length of time of the outage. As the length of the outage increases, the impact of the outage on the business—both quantitatively and qualitatively—also increases. The sooner the business can return to normal, the less overall impact; however, the cost to return to normal is often the inverse of the length of time taken to recover. In other words, the shorter the recovery time frame, the more expensive the cost of the recovery is.

Because the BCP is based on the BIA, the risk practitioner must review the process used to determine the BIA to validate that it is accurate and considers all of the risk factors.

2.16.3 Disaster Recovery Plan

Disaster recovery planning is the recovery of business and IT services following a disaster or incident within a predefined schedule and budget. The time frames for recovery are based on the cost of recovery and length of time that management is willing to accept.

The risk practitioner should review the BCP and DRP to ensure that they are up to date, reflect risk scenarios and business priorities, and have been tested. Testing the plans can uncover any vulnerabilities and provide the experience needed for team members to enact the plan effectively.

2.17 EXCEPTION MANAGEMENT PRACTICES

Having policies, procedures and standards in place is an essential part of operating secure systems and attaining a secure state; however, there may be cases in which an exception to policy, procedure or standard is necessary. Exceptions should only be allowed through a documented, formal process that requires approval of the exception from a senior manager. If exceptions are undocumented and uncontrolled, the level of risk is unknown and may represent a hidden vulnerability. The risk practitioner must ensure that an exception management process is in place and is being followed. After an exception is no longer needed, the exception should be removed.

2.18 IT RISK ASSESSMENT REPORT

As the risk practitioner reaches the conclusion of the IT risk assessment phase, the results of the risk assessment should indicate any gaps between the current risk environment and the desired state of IT risk. The purpose and output of this phase is to provide management with a risk assessment report that documents the risk as well as provides recommendations on how to address any outstanding risk issues. The recommendations made must be justifiable and linked with the results of the risk assessment and findings of the tests and reviews conducted by the risk assessor. Where gaps are found, the risk practitioner must advise management of the areas of noncompliance or unacceptable risk, as well as indicate the severity of the identified issue.

The risk assessment report should document the process used for risk assessment as well as the results of the assessment process. The risk must be documented in a manner that is understandable to management and clearly states the risk levels and priorities.

The risk assessment report may be adjusted according to the needs of the organization and the direction of management, but would normally include the following sections:
• Objectives of the risk assessment process
• Scope and description of the area subject to assessment
• External context and factors affecting risk
• Internal factors or limitations affecting risk assessment
• Risk assessment criteria
• Risk assessment methodology used
• Resources and references used
• Identification of risk, threats and vulnerabilities
• Assumptions used in the risk assessment
• Potential of unknown factors affecting assessment
• Results of risk assessment
• Recommendations and conclusions

The risk assessment should be performed in a consistent manner that supports future risk assessment efforts and would provide predictable results.

2.19 RISK OWNERSHIP AND ACCOUNTABILITY

The completion of the risk assessment phase results in the documentation of all risk and the assessment of the priority of the risk for risk response. Each risk must be linked to an individual who has the responsibility to accept ownership of the risk. The risk owner is tasked with making the decision of what the best response is to the identified risk. The risk owners must be at a level in the organization where they are authorized to make decisions on behalf of the organization and where they can be held accountable for those decisions. To ensure accountability, the ownership of risk must be with an individual, not with a department or the organization as a whole.

2.20 COMMUNICATING THE IT RISK ASSESSMENT RESULTS

The results of the risk assessment should be compiled into a risk assessment report for submission to senior management. The report should list all risk (often in a risk register) and list the seriousness of the risk and the owner of the risk. Wherever possible, the risk practitioner should also provide recommendations as to what action (or inaction) the organization should take in response to the identified risk. These recommendations draw on the experience and knowledge that the risk practitioner has and provide valuable direction and advice for management. They are only recommendations, however, and management can elect to follow or disregard the recommendations as they see fit in the next phase of risk management, the risk response and mitigation phase.

All risk should be noted in the risk assessment report, including issues that may already have been addressed so that there is a record of the detected risk and the actions taken to resolve the risk. This ensures that a control that may have been in place is not removed later, and the risk practitioner will not be suspected of missing the risk during the risk identification and assessment processes.

2.20.1 Risk Reporting Tools and Techniques

There are several excellent commercial tools available for recording and reporting risk, but many organizations use a simple spreadsheet or template for reporting risk. It is important to use a consistent manner to report risk, so that current risk reports can be compared to previous reports and to ensure understandability of the report data. The risk practitioner should ensure that the report is clear, concise and accurate and is free from terminology that could be misunderstood or is subject to misinterpretation.

2.20.2 Updating the Risk Register

Risk management is an ongoing process, and each step of risk identification, assessment, response and monitoring will be repeated on a regular basis. The consistent way to record and track risk is on a risk register. The risk register is a living document that is continuously updated with new data on emerging risk, changes in risk, resolution or completion of a risk response, status updates, or changes in risk ownership and accountability.

At the end of the risk assessment phase, the risk register that was created or updated in the first step (risk identification) should be updated with new information learned in this phase about risk assessment or newly identified risk and responses.

2.21 SUMMARY

During the risk assessment phase, the risk practitioner has a responsibility to assess or determine the severity of each risk facing the organization. Much of this work is based on the results of the risk identification phase, but the risk practitioner should also validate the work of the previous phase and ensure that, as much as possible, all risk is identified, assessed, documented and reported to senior management.

ENDNOTES

[1] International Organization for Standardization (ISO); *IEC 31010:2009: Risk management—Risk assessment techniques*, Switzerland, 2009

[2] ISO/International Electrotechnical Commission; *ISO/IEC 27005:2008: Information technology—Security techniques—Information security risk management*, Switzerland, 2008.

[3] *Op cit* IEC 31010:2009

[4] Panda, Parthajit, "The OCTAVE Approach to Information Security Risk Assessment," *ISACA Journal*, Volume 4, 2009

[5] *Ibid.*

[6] *Ibid.*

[7] *Ibid.*

[8] Kissel, Richard; Kevin Stine; Matthew Scholl; Hart Rossman; Jim Fahlsing; Jessica Gulick; *NIST Special Publication 800-64 Revision 2: System Considerations in the System Development Life Cycle*, National Institute of Standards and Technology (NIST), USA, 2008

Chapter 3: Risk Response and Mitigation

Section One: Overview

Section Two: Contents

Section One: Overview

DOMAIN DEFINITION
Determine risk response options and evaluate their efficiency and effectiveness to manage risk in alignment with business objectives.

LEARNING OBJECTIVES
The objective of this domain is to ensure that the CRISC candidate has the knowledge necessary to:
- List the different risk response options
- Define various parameters for risk response selection
- Explain how residual risk relates to inherent risk, risk appetite and risk tolerance
- Discuss the need for performing a cost-benefit analysis when determining a risk response
- Develop a risk action plan
- Explain the principles of risk ownership
- Leverage understanding of the system development life cycle (SDLC) process to implement IS controls efficiently and effectively
- Understand the need for control maintenance

CRISC EXAM REFERENCE
This domain represents 23 percent of the CRISC exam (approximately 35 questions).

TASK AND KNOWLEDGE STATEMENTS

TASKS
There are seven tasks within this domain that a CRISC candidate must know how to perform. These relate to IT risk identification.

T3.1   Consult with risk owners to select and align recommended risk responses with business objectives and enable informed risk decisions.

T3.2   Consult with, or assist, risk owners on the development of risk action plans to ensure that plans include key elements (e.g., response, cost, target date).

T3.3   Consult on the design and implementation or adjustment of mitigating controls to ensure that the risk is managed to an acceptable level.

T3.4   Ensure that control ownership is assigned in order to establish clear lines of accountability.

T3.5   Assist control owners in developing control procedures and documentation to enable efficient and effective control execution.

T3.6   Update the risk register to reflect changes in risk and management's risk response.

T3.7   Validate that risk responses have been executed according to the risk action plans.

KNOWLEDGE STATEMENTS
The CRISC candidate should be familiar with the task statements relevant to each domain in the CRISC job practice. The tasks are supported by 41 knowledge statements that delineate each of the areas in which the risk practitioner must have a good understanding in order to perform the tasks. Many knowledge statements support tasks that cross domains.

The CRISC candidate should have knowledge of:
1. Laws, regulations, standards and compliance requirements
2. Industry trends and emerging technologies
3. Enterprise systems architecture (e.g., platforms, networks, applications, databases and operating systems)
4. Business goals and objectives
5. Contractual requirements with customers and third-party service providers

6. Threats and vulnerabilities related to:
   - 6.1. Business processes and initiatives
   - 6.2. Third-party management
   - 6.3. Data management
   - 6.4. Hardware, software and appliances
   - 6.5. The system development life cycle (SDLC)
   - 6.6. Project and program management
   - 6.7. Business continuity and disaster recovery management (DRM)
   - 6.8. Management of IT operations
   - 6.9. Emerging technologies
7. Methods to identify risk
8. Risk scenario development tools and techniques
9. Risk identification and classification standards, and frameworks
10. Risk events/incident concepts (e.g., contributing conditions, lessons learned, loss result)
11. Elements of a risk register
12. Risk appetite and tolerance
13. Risk analysis methodologies (quantitative and qualitative)
14. Organizational structures
15. Organizational culture, ethics and behavior
16. Organizational assets (e.g., people, technology, data, trademarks, intellectual property) and business processes, including enterprise risk management (ERM)
17. Organizational policies and standards
18. Business process review tools and techniques
19. Analysis techniques (e.g., root cause, gap, cost-benefit, return on investment [ROI])
20. Capability assessment models and improvement techniques and strategies
21. Data analysis, validation and aggregation techniques (e.g., trend analysis, modeling)
22. Data collection and extraction tools and techniques
23. Principles of risk and control ownership
24. Characteristics of inherent and residual risk
25. Exception management practices
26. Risk assessment standards, frameworks and techniques
27. Risk response options (i.e., accept, mitigate, avoid, transfer) and criteria for selection
28. Information security concepts and principles, including confidentiality, integrity and availability of information
29. Systems control design and implementation, including testing methodologies and practices
30. The impact of emerging technologies on design and implementation of controls
31. Requirements, principles, and practices for educating and training on risk and control activities
32. Key risk indicators (KRIs)
33. Risk monitoring standards and frameworks
34. Risk monitoring tools and techniques
35. Risk reporting tools and techniques
36. IT risk management best practices
37. Key performance indicator (KPIs)
38. Control types, standards, and frameworks
39. Control monitoring and reporting tools and techniques
40. Control assessment types (e.g., self-assessments, audits, vulnerability assessments, penetration tests, third-party assurance)
41. Control activities, objectives, practices and metrics related to:
    - 41.1. Business processes
    - 41.2. Information security, including technology certification and accreditation practices
    - 41.3. Third-party management, including service delivery
    - 41.4. Data management
    - 41.5. The system development life cycle (SDLC)
    - 41.6. Project and program management
    - 41.7. Business continuity and disaster recovery management (DRM)
    - 41.8. IT operations management
    - 41.9. The information systems architecture (e.g., platforms, networks, applications, databases and operating systems)

SELF-ASSESSMENT QUESTIONS

3-1   When a risk cannot be sufficiently mitigated through manual or automatic controls, which of the following options will **BEST** protect the enterprise from the potential financial impact of the risk?

    A. Insuring against the risk
    B. Updating the IT risk register
    C. Improving staff training in the risk area
    D. Outsourcing the related business process to a third party

3-2   To be effective, risk mitigation **MUST**:

    A. minimize the residual risk.
    B. minimize the inherent risk.
    C. reduce the frequency of a threat.
    D. reduce the impact of a threat.

3-3   The **BEST** control to prevent unauthorized access to an enterprise's information is user:

    A. accountability.
    B. authentication.
    C. identification.
    D. access rules.

3-4   Which of the following controls **BEST** protects an enterprise from unauthorized individuals gaining access to sensitive information?

    A. Using a challenge response system
    B. Forcing periodic password changes
    C. Monitoring and recording unsuccessful logon attempts
    D. Providing access on a need-to-know basis

3-5   Which of the following defenses is **BEST** to use against phishing attacks?

    A. An intrusion detection system (IDS)
    B. Spam filters
    C. End-user awareness
    D. Application hardening

3-6   When responding to an identified risk event, the **MOST** important stakeholders involved in reviewing risk response options to an IT risk are the:

    A. information security managers.
    B. internal auditors.
    C. incident response team members.
    D. business managers.

3-7   Which of the following choices should be considered **FIRST** when designing information system controls?

   A. The organizational strategic plan
   B. The existing IT environment
   C. The present IT budget
   D. The IT strategic plan

3-8   What risk elements **MUST** be known in order to accurately calculate residual risk?

   A. Threats and vulnerabilities
   B. Inherent risk and control risk
   C. Compliance risk and reputation
   D. Risk governance and risk response

ANSWERS TO SELF-ASSESSMENT QUESTIONS
Correct answers are shown in **bold**.

3-1   **A.   An insurance policy can compensate the enterprise monetarily for the impact of the risk by transferring the risk to the insurance company.**
  B.   Updating the risk register (with lower values for impact and probability) will not actually change the risk, only management's perception of it.
  C.   Staff capacity to detect or mitigate the risk may potentially reduce the financial impact, but insurance allows for the risk to be completely mitigated.
  D.   Outsourcing the process containing the risk does not necessarily remove or change the risk.

3-2   **A.   The objective of risk reduction is to reduce the residual risk to levels below the enterprise's risk tolerance level.**
  B.   The inherent risk of a process is a given and cannot be affected by risk reduction/risk mitigation efforts.
  C.   Risk reduction efforts can focus on either avoiding the frequency of the risk or reducing the impact of a risk.
  D.   Risk reduction efforts can focus on either avoiding the frequency of the risk or reducing the impact of a risk.

3-3   A.   User accountability does not prevent unauthorized access; it maps a given activity or event back to the responsible party.
  B.   Authentication verifies the user's identity and the right to access information according to the access rules.
  C.   User identification without authentication does not grant access.
  D.   Access rules without identification and authentication do not grant access.

3-4   A.   Verifying the user's identification through a challenge response does not completely address the issue of access risk if access was not appropriately designed in the first place.
  B.   Forcing users to change their passwords does not guarantee that access control is appropriately assigned.
  C.   Monitoring unsuccessful access logon attempts does not address the risk of appropriate access rights.
  D.   Physical or logical system access should be assigned on a need-to-know basis, when there is a legitimate business requirement based on least privilege and segregation of duties (SoD).

3-5   A.   An intrusion detection system (IDS) does not protect against phishing attacks because phishing attacks usually do not have the same patterns or unique signatures.
  B.   While certain highly specialized spam filters can reduce the number of phishing emails that reach their addressees' inboxes, they are not as effective in addressing phishing attacks as end-user awareness.
  C.   Phishing attacks are a type of social engineering attack and are best defended by end-user awareness training.
  D.   Application hardening does not protect against phishing attacks because phishing attacks generally use email as the attack vector, with the end user, not the application, as the vulnerable point.

3-6   A.   Information security managers may best understand the technical tactical situation, but business managers are accountable for managing the associated risk and will determine what actions to take based on the information provided by others, which includes collaboration with, and support from, IT security managers.
  B.   This is not a function of internal audit. Business managers set priorities, possibly consulting with other parties, which may include internal audit.
  C.   The incident response team must ensure open communication to management and stakeholders to ensure that business managers/leaders understand the associated risk and are provided enough information to make informed risk-based decisions.
  D.   Business managers are accountable for managing the associated risk and will determine what actions to take based on the information provided by others.

3-7 **A. Review of the enterprise's strategic plan is the first step in designing effective IS controls that would fit the enterprise's long-term plans.**

   B. Review of the existing IT environment, although useful and necessary, is not the first task that needs to be undertaken.

   C. The present IT budget is one of the components of the strategic plan.

   D. The IT strategic plan exists to support the enterprise's strategic plan.

3-8 A. Threats and vulnerabilities are elements of inherent risk. They do not accurately calculate residual risk.

   B. Inherent risk (threats × vulnerabilities) multiplied by control risk is the formula to calculate residual risk.

   C. Compliance risk is the current and prospective risk to earnings or capital arising from violations of, or nonconformance with, laws, rules, regulations, prescribed practices, internal policies and procedures, or ethical standards. Compliance risk can lead to reputational damage.

   D. Risk governance and risk response are risk domains, not risk elements, to calculate residual risk.

NOTE: For more self-assessment questions, you may also want to obtain a copy of the *CRISC™ Review Questions, Answers & Explanations Manual 2015*, which consists of 400 multiple-choice study questions, answers and explanations, and the *CRISC™ Review Questions, Answers & Explanations Manual 2015 Supplement*, which consists of 100 new multiple-choice study questions, answers and explanations.

SUGGESTED RESOURCES FOR FURTHER STUDY

In addition to the resources cited throughout this manual, the following resources are suggested for further study in this domain (publications in **bold** are stocked in the ISACA Bookstore):

Abkowitz, Mark D.; *Operational Risk Management: A Case Study Approach to Effective Planning and Response*, John Wiley & Sons, USA, 2008

International Organization for Standardization (ISO)/International Electrotechnical Commission (IEC); *ISO/IEC 27001:2013 Information technology—Security techniques—Information security management systems—Requirements*, Switzerland, 2013

ISO/IEC; *ISO/IEC 27005:2011—Information Security Risk Management*, Switzerland, 2011

ISO/IEC; *ISO/IEC 31000:2009 Risk management—Principles and Guidelines*, Switzerland, 2009

ISACA, *COBIT® 5 for Risk*, USA, 2013, *www.isaca.org/cobit*

ISACA, *The Risk IT Framework*, USA, 2009

ISACA, *The Risk IT Practitioner Guide*, USA, 2009

The Open Web Application Security Project, www.owasp.org

Section Two: Content

3.0 OVERVIEW

The risk response phase of risk management, shown in **exhibit 3.1**, requires management to make the decisions regarding the correct way to respond to and address risk. The risk response decision is based on the information provided in the earlier steps of risk identification and risk assessment, but is balanced with the constraints placed on the organization through budget, time, resources, strategic plans, regulations and customer expectations. Management must be prepared to justify its risk response decision and provide a road map to implementing the changes that have been decided on according to a reasonable schedule. The risk response must ensure that business operations are protected but not unduly impaired or impacted by controls that are put in place to address risk.

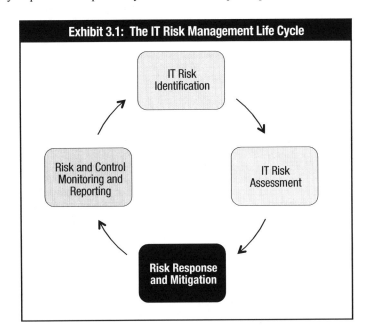

Exhibit 3.1: The IT Risk Management Life Cycle

3.1 ALIGNING RISK RESPONSE WITH BUSINESS OBJECTIVES

The risk assessment report and risk register document the risk as identified by the risk practitioner during the IT risk identification and IT risk assessment phases, and they should indicate the assessed level of, or priority of, each risk. Management is responsible for evaluating and responding to the recommendations included in the report. The recommendations in the report are guidelines, and several recommendations may be provided. Management is responsible for determining the best response to the risk and for developing an action plan and implementation strategy to resolve any outstanding risk in a prudent and thoughtful manner based on the level of risk acceptance and tolerance of the organization. Management must always be aware of the drivers for risk management, such as compliance with regulations and the need to support and align the risk response with business priorities and objectives. The organization exists to meet its mission. A risk response that negatively impacts the ability of the organization to meet its mission must be carefully considered, but balanced against the need to meet regulatory requirements and customer expectations.

In some cases, the organization will require the development and approval of several individual projects and initiatives to complete its entire risk response strategy. These may be arranged into one program and scheduled based on the dependencies between the various projects and the priorities for risk mitigation.

3.2 RISK RESPONSE OPTIONS ~~Risk Treatment~~

The purpose of defining a risk response is to bring risk in line with the defined risk appetite for the enterprise. This risk response evaluation is not a one-time effort; rather, it is part of the risk management process cycle. When a risk analysis of all identified risk scenarios has shown—after weighing risk versus potential return—that a risk is not aligned with the defined risk appetite and tolerance levels, a response is required.

There are four commonly accepted options for risk response:
- Risk acceptance
- Risk mitigation
- Risk avoidance
- Risk transfer

3.2.1 Risk Acceptance

The choice to accept risk is a conscious decision made by senior management to recognize the existence of risk and knowingly decide to allow (assume) the risk to remain without (further) mitigation. Therefore, management is responsible for the impact of a risk event should it occur. This decision is made according to the risk acceptance level set by senior management, but with some consideration of risk tolerance. Risk acceptance is the amount of risk that senior management has determined is within acceptable or permissible bounds. However, risk acceptance is the not the same as risk ignorance.

Risk ignorance is the failure to identify or acknowledge risk, or it is a decision to blindly accept risk without knowing or acknowledging what the risk level really is. A failure to properly assess the risk or to understand the level of risk may be an example of a lack of due care or action by management toward protecting the assets of the organization.

Risk tolerance is used as an exceptional circumstance where the level of risk may exceed the risk acceptance boundary set by senior management, but the decision is made to accept the risk anyway. This could occur because no controls are available, or the cost of the controls would outweigh the benefits they provide. A risk tolerance exception must be approved by management with appropriate levels of authority to accept the risk on behalf of the organization.

Some examples of risk acceptance are as follows:
- It is predicted that a certain project will not deliver the required business functionality by the planned delivery date. Management may decide to accept the risk and proceed with the project.
- A particular risk is assessed to be extremely rare but very important (catastrophic), and approaches to reduce it are prohibitive. Management may decide to accept this risk.

Self-insurance, or deciding to absorb the potential costs of an incident, is another form of risk acceptance. However, this manages only magnitude of the loss and has no impact on frequency.

Risk acceptance is often based on poorly calculated risk levels, and many organizations have found that the level of risk they had intended to accept was far less than the amount of damage sustained after the incident occurred. The level of damage is difficult to accurately determine and is often underestimated. Therefore, a review of similar incidents at other organizations and the resulting impact on those firms is often a good guideline to indicate true incident costs.

A principle of risk acceptance is also addressed in ISO/IEC 27001:2013 *Information technology—Security techniques – Information security management systems – Requirements*: "The organization shall define and apply an information security risk treatment process to obtain risk owner's approval for the information security risk treatment plan and acceptance of the residual information security risks."[1] This indicates the need to know the expected residual risk and ensure that the risk treatment plan does meet the expectations of the risk owner.

The level of risk and impact can change dramatically as assets increase or decrease in value, new threats emerge, attackers gain additional skills or motivation, vulnerabilities are discovered, the existing controls become weaker, or the operations team becomes less diligent. Therefore, regular reviews of risk the organization has chosen to accept should be conducted.

3.2.2 Risk Mitigation

Risk mitigation means that action is taken to reduce the frequency and/or impact of a risk. The greater the area affected by the risk or the speed at which an incident may expand affects risk mitigation choices. Risk may require mitigation through several controls until it reaches the level of risk acceptance or, in an exceptional case, risk tolerance.

Some examples of risk mitigation are:
• Strengthening overall risk management practices, such as implementing sufficiently mature risk management processes
• Deploying new technical, management or operational controls that reduce either the likelihood or the impact of an adverse event
• Installing a new access control system
• Implementing policies or operational procedures
• Developing an effective incident response and business continuity plan (BCP)
• Using compensating controls

3.2.3 Risk Avoidance

Risk avoidance means exiting the activities or conditions that give rise to risk. Risk avoidance applies when no other risk response is adequate. This is the case when:
• There is no other cost-effective response that can succeed in reducing the frequency and impact below the defined thresholds for risk appetite.
• The risk cannot be shared or transferred.
• The exposure level is deemed unacceptable by management.

Some IT-related examples of risk avoidance include:
• Relocating a data center away from a region with significant natural hazards
• Declining to engage in a very large project when the business case shows a notable risk of failure
• Declining to engage in a project that would build on obsolete and convoluted systems because there is no acceptable degree of confidence that the project will deliver anything workable
• Deciding not to use a certain technology or software package because it would prevent future expansion

The risk practitioner should support the organization through revising risk evaluations and providing supporting data for management to revisit earlier decisions and potentially make changes in strategy or strategic plans.

3.2.4 Risk Sharing/Transfer

Risk transfer is a decision to reduce loss through sharing the risk of loss with another organization. This is often done through purchasing insurance. For example, an organization may purchase insurance that covers fire damage and pay an annual insurance premium. If the organization never has a fire, the money spent on insurance premiums is lost profit; if it does have a claim, the benefit provided by insurance may allow the organization to recover and continue operations more quickly than if it had no insurance.

Another example of risk transfer is to bid on a large project in partnership with another organization. This allows the organization to accept the risk of winning a large contract even though they cannot meet all the requirements of the contract themselves. This means that the organization also has to split the profit of the contract with their business partners. Therefore, by transferring a portion of the risk, they also lose a portion of the benefit.

When the organization chooses to transfer risk, this decision must be reviewed on a regular basis. For example, an organization should ensure that the current amount of insurance is adequate to cover losses and that the organization is compliant with the terms and conditions of the coverage.

3.3 ANALYSIS TECHNIQUES

The determination of the best risk response can be based on one of several analysis techniques. The risk response chosen by the management of the organization is often based on the evaluation of several possible response options. The management of the organization should consider several factors in selecting a response, such as:
• The priority of the risk as indicated in the risk assessment report
• The recommended controls from the risk assessment report
• Any other response alternatives that are suggested through further analysis
• The cost of the various response options, including:
  – Acquisition cost
  – Training cost
  – Impact of productivity
  – Maintenance and licensing costs
• Requirements for compliance with regulations or legislation

- Alignment of the response option with the strategy of the organization
- Possibility of integrating the response with other organizational initiatives
- Compatibility with other controls in place
- Time, resources and budget available

The selection of the appropriate response is based primarily on the calculation of value obtained for cost incurred. In other words, does the cost of implementing a specific risk response provide enough value to the organization for it to be a wise decision? In the end, the decision may be to accept the risk or to respond to the risk through a combination of managerial, technical and physical controls.

3.3.1 Cost-benefit Analysis
Cost-benefit analysis is used to justify the expense associated with the implementation of controls. The expenditure on a control cannot be justified if the benefit realized from the control is less than the cost. There are several factors that must be included in calculating the total cost of the control:
- Cost of acquisition
  - Evaluation of solutions
  - Cost of the control
  - Cost of training
  - Cost to rearchitect systems
- Ongoing cost of maintenance
  - License costs
  - Cost of staff to monitor and report on control
  - Impact on productivity/performance
  - Cost of support and technical assistance
- Cost to remove/replace control

Likewise, the benefit realized from the control must consider several factors:
- Reduced cost of risk event
- Reduced liability
- Reduced insurance premiums
- Increased customer confidence
- Increased shareholder confidence
- Trust from financial backers
- Faster recovery
- Better employee relations (safety)

The costs and benefits may be calculated using both qualitative and quantitative measures. The impact is often related to the length of the outage, the frequency of the outage (i.e., customers may more readily forgive one incident than frequent problems), and scope and publicity of the incident.

3.3.2 Return on Investment
Many organizations use return on investment (ROI) as a method of justifying an investment in a project, tool or new venture. In ROI calculations, an investment would be expected to pay for itself within a set time period. A new computer system, for example, would be expected to pay for itself through better productivity, lower number of staff required or increased sales.

The ROI based on the implementation of a control is often difficult to determine. The cost of the control and the ROI of the control may need to be spread over several years, and it is hard to predict the likelihood of a successful attack. Some organizations have used the term return on security investment (ROSI) to refer to the ROI specifically in relation to the payback for security controls; however, the ROSI can be difficult to calculate because it can be impossible to forecast what the true benefit of a control would have been depending on whether an adverse event occurred. In determining ROI or ROSI, the organization is trying to forecast the likelihood and impact of an incident and deciding what is an adequate level of protection. Like the purchase of insurance, this is a cost that may or may not provide a direct benefit in the future. In one way, the amount of security an organization decides to implement is dependent on their appetite for risk and perception of the probability of exposure.

3.4 VULNERABILITIES ASSOCIATED WITH NEW CONTROLS

Implementing a new or modified control may provide benefit to the organization, but it may also introduce new vulnerabilities that may pose a risk. For example, an access control system may protect an organization from unauthorized access, but will also frequently affect normal users who forget passwords or lose access cards. This will result in the access control system causing a denial of service to authorized personnel and result in more calls for technical assistance, delays in processing and user frustration.

3.5 DEVELOPING A RISK ACTION PLAN

There are several responses to risk that may be considered, and the risk practitioner plays a consultative role in assisting risk owners with deciding on the correct response to a risk (i.e., to accept, avoid, transfer or mitigate a risk). The ultimate decision on risk response is the responsibility of the risk owner, but the risk practitioner can provide advice on technologies, policies, procedures, control effectiveness and leveraging of existing controls.

The decision plan to implement a control is often based on factors such as:
• Current risk level
• Regulations
• Ongoing projects
• Strategic plans
• Budget
• Availability of staff
• Public pressure
• Actions of competitors

The risk response process, as shown in **exhibit 3.2**, includes the following concepts:
1. Risk scenarios (from risk identification) drive the risk analysis and assessment.
2. Risk analysis (from risk assessment) leads to the documentation (mapping) of risk.
3. Risk response is determined by the risk appetite of the organization. If the risk exceeds the appetite, then a risk response to address the risk must be considered.
4. The risk response options are considered along with the risk response parameters to determine the best available risk response.
5. The selected risk response is documented.
6. The risk responses are prioritized according to the current risk environment and the cost-benefit of risk mitigation.
7. A risk action plan is created to manage the risk response projects.

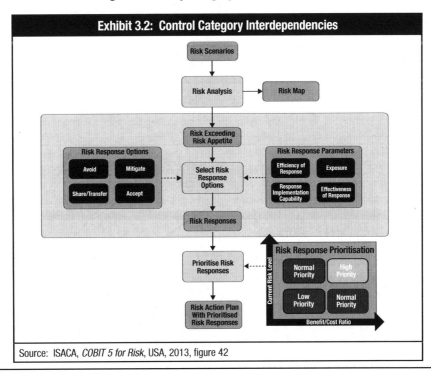

Exhibit 3.2: Control Category Interdependencies

Source: ISACA, *COBIT 5 for Risk*, USA, 2013, figure 42

The risk response may also be influenced by several risk parameters:
• The decision to implement a certain control may be subject to the strategic direction and technological direction of the organization, which can affect the following:
 – The choice of controls available
 – The schedule for deployment (there may be little value in implementing a new control for a technology that is about to be replaced)
• There are often several controls available and the organization (with the advice of the risk practitioner) may need to select which control to implement from among the available choices.
 – Is a more expensive control better than an inexpensive one?
 – Is a control that provides many functions better that one that is specialized for one specific purpose?

These decisions can be difficult to make and will often require some method of comparing control options (perhaps in a matrix that compares cost, function, support, etc.) to determine which control to select.

The choice of which control to implement (managerial, technical or physical, or a combination) includes reviewing the efficiency of the proposed control:
• Would it be effective?
• Would it provide a satisfactory ROI?
• Does the organization have sufficient skill to implement, configure and maintain the control?
• Is there sufficient budget and time to implement the control?
• How much would the control cost to operate on an annual basis, or what would the impact of the control be on productivity?

Each control should be effective; measureable; provide accurate, timely information; and, when possible, be aligned with the culture, technology, budget and strategy of the organization.

The risk action plan should be run as a project. This means that it must have a defined start and end date. The end date is often used to determine the critical path or the elements of the project that may have a direct impact on whether the project can meet the end date. A change in the delivery of any project element that is on the critical path affects the delivery of the entire project. For example, a project that does not receive its equipment from the supplier on time may not be able to meet the scheduled project dates. This represents a risk to the project. Through experience and careful evaluation, the risk practitioner can advise the risk owner on the feasibility of project dates, the expected workload associated with the project, the costs of the project and the overall success of the project according to risk management and business goals.

3.6 BUSINESS PROCESS REVIEW TOOLS AND TECHNIQUES

A business process review examines the effectiveness and efficiency of an organization in meeting its goals and objectives. When conducting a risk response, the organization should review its business processes to ensure that the correct solution is selected to integrate and work effectively with the existing environment, or whether the environment should be reengineered to improve its efficiency and success in meeting organizational objectives.

A business process review requires input from knowledgeable representatives from all affected departments within the organization, and it may also bring in external experts who can provide advice and assistance. The purpose of the business process review is to identify ways to:
• Identify problems or issues with the current process
• Gather information toward improving processes
• Prepare a road map to implement required changes
• Assign responsibility for projects
• Schedule individual projects according to priority
• Monitor project progress and deliverables
• Review and obtain feedback on project results

The steps of a business process review are:
1. Document and evaluate current business processes
    a. List critical processes, supply chains and services
    b. Document current business processes and risk
        i. Review documentation
        ii. Interview management, users and other stakeholders
        iii. Observe actual processes
        iv. Validate with business representatives
            1. Validate training and skills required of personnel
    c. Document current issues and problems
    d. Baseline other organizations
    e. Work with team members to discover potential solutions and improvements
2. Identify potential changes
    a. Use focus groups and workshops to determine process improvements
    b. Validate proposed changes with management
3. Schedule and implement changes
    a. Design changes
    b. Identify dependencies
    c. Communicate the schedule of changes
4. Feedback and evaluation
    a. Measure operational efficiencies
        i. Customer satisfaction
        ii. User satisfaction
        iii. Improvements in productivity/quality
        iv. Feedback to improve processes

3.7 CONTROL DESIGN AND IMPLEMENTATION

Risk mitigation is accomplished through the use of controls. Controls may be proactive, meaning that they attempt to prevent an incident. Or, controls may be reactive, meaning that they allow the detection, containment and recovery from an incident. Proactive controls are often called safeguards, and reactive controls are known as countermeasures. For example, a sign that warns a person about a dangerous condition is a safeguard, whereas a fire extinguisher or sprinkler system is a countermeasure.

Every organization has some controls in place, and the risk assessment should document those controls and calculate their effectiveness in mitigating risk. In some cases, the controls may be sufficient, whereas in others, the controls may need adjustment or replacement. An effective control is one that prevents, detects and/or contains an incident and enables recovery from a risk event.

> **The risk practitioner will provide advice on the selection, design, implementation, testing and operation of the controls.**

It is common for an organization to have some situations where the controls currently in place are not sufficient to adequately protect the organization. In most cases, this requires the adjustment of the current controls or the implementation of new controls. However, it may not be feasible to reduce the risk to an acceptable level by either adjusting or implementing controls due to reasons such as cost, job requirements or availability of controls. An example of this could be found in a small organization when an individual is given administrator rights on a system and there is not adequate segregation of duties (SoD). In this case, it may not be feasible to implement a new or enhanced control; some personnel need administrator rights to perform their jobs, and the risk cannot justify the cost of hiring new staff to address SoD. Therefore, the risk owner may consider implementing compensating controls to reduce the risk. Compensating controls address the weaknesses in the existing controls through concepts such as layered defense, increased supervision, procedural controls, or increased audits and logging of system activity. These measures will work to compensate for the risk that could not be addressed in other ways.

Controls may be grouped into managerial, technical or physical controls, and within each of those groups of controls are various types of controls that can be used, such as directive, deterrent, preventive, detective, corrective, recovery and compensating controls. Controls are discussed in more detail in chapter 2, IT Risk Assessment. An example of a control matrix is shown in **exhibit 3.3**. Many controls may fit into more than one category.

Exhibit 3.3: Control Matrix			
	Managerial	**Technical**	**Physical**
Directive	Policy	Notification that this is a private computer system	"No Trespassing" sign
Deterrent	Disciplinary policy	Warning banner on login	"Beware of Dog" sign
Preventive	User registration process	Login screen	Fence
Detective	Audit	Intrusion detection system (IDS)	Motion sensor
Corrective	Remove access	Network isolation	Close fire doors
Recovery	Revised business processes	Restore from backups	Rebuild damaged building
Compensating	Separation of duties (SoD)	Two-factor authentication	Dual control operations

The interaction between the control types is shown in **exhibit 3.4**.

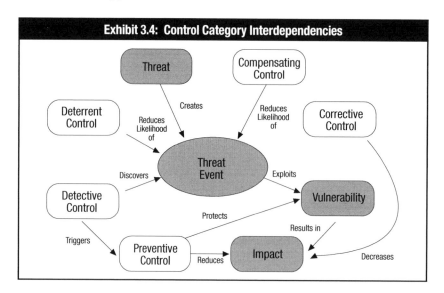

3.7.1 Control Standards and Frameworks

The selection of controls requires the evaluation and implementation of the right control in the right way. Based on data collected through an analysis method (e.g., cost-benefit analysis, ROI, etc.), management will decide on the best available control, or group of controls, to mitigate a specific risk. However, a poorly implemented control may pose a significant risk to the organization by creating a false sense of security or leading to a denial of service if the control does not function correctly. The implementation of a technical control requires that the control is surrounded by proper procedures, the personnel that operate it are adequately trained, a person is assigned ownership of the control (often the person who owns the risk), and the control is monitored and tested to ensure its correct operation and effectiveness.

Many industries have standards that may be used as a benchmark for security across the industry sector. One example is the Payment Card Industry Data Security Standard (PCI DSS), which is used as a standard for all organizations that process payment cards (e.g., debit cards, credit cards, etc.). This is an example of an industry standard, but compliance is not required by law. Such standards, and the frameworks that implement those standards, are found in the health care, accounting, audit and telecommunications industries. In some regulated industries, regulations require compliance with a standard, such as the electrical power industry. To meet the requirements of the standard, a framework is often used to describe how an organization can achieve compliance.

A control framework is defined as a set of fundamental controls that facilitates the discharge of business process owner responsibilities to prevent financial or information loss in an enterprise. Therefore, it can be seen as the implementation of controls intended to support and protect business operations and preserve asset value.

3.7.2 Administrative, Technical and Physical Controls
Controls are often divided into three groups: administrative, technical and physical, as shown in **exhibit 3.5**.

Exhibit 3.5: Control Methods	
Category	**Description**
Administrative controls	Administrative controls are managerial in nature and are related to the oversight, reporting, procedures and operations of a process. These include controls such as policy, procedures, balancing, employee development and compliance reporting.
Technical controls	Technical controls are sometimes known as logical controls and are provided through the use of a technology, equipment or device. Examples of technical controls include firewalls, network or host-based intrusion detection systems (IDSs), passwords and antivirus software. A technical control requires proper management (administrative) controls to operate correctly.
Physical controls	Physical controls are locks, fences, closed-circuit TV (CCTV) and such devices that are installed to physically restrict access to a facility or hardware. Physical controls require maintenance, monitoring and the ability to assess and react to an alert should a physical control indicate a problem.

3.8 CONTROL MONITORING AND EFFECTIVENESS
To support the ability to monitor and report on risk, the risk practitioner should ensure that processes, logs and audit hooks are placed into the control framework. This allows for the monitoring and evaluation of controls. As controls are designed, implemented and operated, the risk practitioner should ensure that logs are enabled, controls are able to be tested and regular reporting procedures are developed.

3.8.1 Control Monitoring and Reporting Tools and Techniques
The actual monitoring and reporting of controls should be performed in the risk monitoring phase of risk management; however, the risk practitioner should ensure that the capability to monitor a control and to support monitoring systems is addressed in control design. If the organization is using a managed security service provider (MSSP) or a security event and incident management (SEIM) system, the ability to capture data, and the notification to the operations staff on the deployment of the system, are necessary.

3.9 CHARACTERISTICS OF INHERENT AND RESIDUAL RISK
Some degree of risk is unavoidable in business; however, some business processes have a higher level of inherent risk than others. Risk is inherent in everything; however, the degree of risk varies from one activity, product or service to another. The risk practitioner should understand the inherent risk, be able to assess the current risk, and if necessary, mitigate the risk to reduce it to an acceptable level of residual risk.

3.9.1 Inherent Risk
Inherent risk is the risk level or exposure without taking into account the actions that management has taken or might take (e.g., implementing controls). When assessing risk, the risk practitioner should examine the current risk profile of the organization including the inherent risk, the levels of residual risk, and the correct operation and monitoring of the controls in place.

An area that has a higher level of inherent risk may need additional controls to reduce the level of risk to an acceptable level. An example is an area that handles cash or other financial transactions. Such areas often require stronger physical controls, more monitoring, SoD and background checks for employees.

3.9.2 Residual Risk

The objective of mitigating controls is to reduce the level of risk. The level of risk that remains following the implementation of a control is the residual (remaining) risk.

> **Residual Risk = Total Risk – Control Effectiveness**

Through the implementation of controls, the residual risk should be reduced to a level that is less than or equal to acceptable risk. Acceptable risk is the level of risk that management is willing to bear or accept. When controls have been put in place, the combined effect of the controls should be a calculation of remaining risk that is within acceptable bounds.

3.10 CONTROL ACTIVITIES, OBJECTIVES, PRACTICES AND METRICS

During the risk response and mitigation phase of risk management, controls are chosen to address or respond to risk. The objective is that by the end of this phase, IT risk will be within acceptable risk levels. This requires the measurement and monitoring of risk. This phase will also support the next phase, risk and control monitoring and reporting, by putting into place the mechanisms and ability to measure risk and monitor the effectiveness of the controls. The use of controls is often compared to industry standards, benchmarks and recognized good practices.

3.10.1 Business Processes

The management of risk and implementation of controls should not unduly affect business operations. The risk management effort must consider the impact of controls on the ability of the business to meet its objectives and the ability of the users to accomplish their tasks in a simple, logical manner.

When a business process involves sensitive information, such as personal health information (PHI), the process must be designed to protect the information from unauthorized disclosure, alteration or deletion. The sensitive data may be masked or hidden from users who do not have a need to know, or may be read-only for users who do not have authority to alter the data. Building these types of controls into the application is a part of effective and preventive risk management.

The risk response should ensure compliance with standards of risk management such as local regulation (e.g., the US Health Insurance Portability and Accountability Act [HIPAA]), international standards (e.g., ISO 27001), industry standards (e.g., PCI DSS) and internal standards (e.g., policy, culture and procedures of the organization).

The result of the risk response may also lead to accreditation of the organization as compliant with an international standard such as ISO/IEC 27001.

3.10.2 Information Security

Risk response will ensure that technology used in the organization is adequately protected, secure and reliable. The use of new technology will be supported through a proactive risk assessment and mitigation program that evaluates the risk associated with new technology and, where possible, provides advice on how to deploy and use the new technology within acceptable risk boundaries. The deployment of new technology includes:
• Training for users and administrators
• Creation of policies and procedures
• Inclusion of such systems in backups and BCPs
• Assignment of ownership for the risk
• Consent of information owners for technology that may handle sensitive information
• Review of legal or regulatory requirements
• Assignment of responsibility for monitoring and reporting on the proper use of such technology

Legacy systems require special attention because they frequently have been built with attention to supporting a business function, but not to address security issues such as buffer overflows or current threats. Legacy systems may also be reaching the end of their operational life and are, therefore, more susceptible to failure. The risk response options for legacy systems may be very limited because the cost of replacing or upgrading such systems is simply not feasible. Compensating controls—additional backups, spare parts, cross-training, documentation and increased monitoring of systems performance—may be needed.

Every system is the responsibility of a system owner. The owner of the system is usually a senior manager in the department for which the system was built. For example, the owner of financial data, and the financial systems, of the organization is often the chief financial officer (CFO). The CFO is responsible for the proper use and operation of the financial system and usually must pay for and approve the implementation of, changes to, upgrades to and removal of the system. The CFO engages the IT department (or an external supplier) to manage and operate the financial systems on the behalf of the finance department. The ownership and responsibility for the correct operation of the system and the protection of the data does not transfer to the outsource supplier or IT department. It remains with the CFO except in a rare circumstance in which the supplier legally accepts responsibility for the protection.

The problem with system ownership being within the business is that an organization may have many system owners from various departments, making a coordinated way to manage, oversee and ensure consistent operation of the systems difficult. An information system that supports the sales department may operate in a considerably different cultural environment than an information system that supports the finance department. This difference in culture may affect the way the system is managed and protected. If the protection for each system is the sole responsibility of the owner of that system, then significant differences can exist in how the security and risk for each system is enforced. A breach or vulnerability in any one system is a risk to the entire organization because an organization is connected via networks, information sharing and system dependencies. Therefore, the organization must take an enterprisewide approach to security to ensure consistency, reliable and secure operations, and integrated risk management. This is done through two common methods: change control and certification and accreditation (C&A).

Change Control
Change control is often managed through a change control board (CCB) or other such committee or mechanism. A CCB is comprised of representatives from several business departments and is responsible for overseeing all information systems operations and approving changes to those systems. This provides a communications channel between the business units and the IT department. As changes are requested or patches are released from vendors, the CCB reviews the change requests to ensure that:
• The change request does not unknowingly affect risk or security.
• The change is formally requested, approved and documented.
• The change is scheduled at a time convenient for the business and IT.
• All stakeholders affected by the change are advised of the change.

This also provides a balance between the need to allow change and the need to preserve system reliability and stability.

Certification and Accreditation
C&A is the process of objective management and acceptance of risk associated with the installation and operation of information systems. The purpose of C&A is to provide an organizationwide approach to the management of IT risk by ensuring that all information systems are subject to review and approved for operation. Certification is the process of reviewing information systems with regard to their secure design, development, testing, deployment and operations. The certifier examines the technical and nontechnical aspects of system operations to ensure that the risk associated with the system has been identified and adequately mitigated. The certifier examines the risk assessments, system documentation and testing to verify that the system will operate within an acceptable level of risk, and provides a report to an authorizing official (AO), recommending whether to accredit the system for operation. Certification should be done in parallel with the system development life cycle (SDLC) to ensure that any unacceptable risk is identified promptly and addressed to system and project management. This should permit the secure development of the project and minimize the risk of a system being developed that could pose an unacceptable level of risk to the organization.

Accreditation is the official, formal decision by a senior manager (the AO) to approve or authorize an information system to operate. This decision is the explicit acceptance of risk by the AO, where the AO knowingly and consciously evaluates the risk associated with the operation of an information system and approves the system for operation. The AO accepts the risk on behalf of the organization (its assets, mission and individuals) as well as the risk to other agencies, organizations and the nation. In some cases, if the needs of the business require it or the risk is minimal, the system may be approved to operate only under certain restrictions. The system owner is then authorized to operate the system according to the restrictions and for the time period granted in the accreditation. Ideally, this process should never fail to approve a system because any issues that would have prevented authorization should have been identified by the certifier and resolved by the system and project owner prior to the accreditation decision.

Asset Inventory and Documentation

An important part of managing risk is the identification and inventory of the assets of the organization. As changes are made to systems and new equipment is installed or older equipment is retired, the asset inventory must be updated to reflect what assets, systems and equipment the organization is currently using. The asset inventory should list all equipment, the location and owner of the equipment and other data required for maintenance, insurance and warranty purposes.

Configuration Management

As devices and systems are installed, they must be configured to support communications, interfaces with other systems and secure operations. This often requires the hardening of systems to disable ports, services and protocols that are not required to be operational on the system. Hardening a system or device reduces the attack vectors or attack surface that could be used as a potential point of compromise. A hardened system is one that has all unnecessary functionality disabled and does not store any sensitive data that are not immediately needed to support a business operation. For example, a web server should be installed onto a secure host, known as a bastion host, that has very little functionality and should never contain credit card data or other sensitive information. A common error is storing such sensitive information on the web server for convenience, only to have these data stolen by an attacker who compromised the web application.

The rules (i.e., configuration) that control the operation of devices, such as a firewall, must also be backed up and available in case of equipment failure. By having a backup of the configuration, the ability to recover the system is facilitated and often much faster.

3.10.3 Third-party Management

The risk practitioner should address the risk that arises when an organization outsources business functions, IT services and data management. The ownership of the data and business processes remains with the organization that is doing the outsourcing. This creates a legal liability that may be difficult for the outsourcing organization to manage because most of the day-to-day operations, staff and procedures are outside of the organization's direct control.

When the management of data is outsourced, the outsourcing organization must ensure that the security requirements and regulations for handling the information have been written into the outsourcing agreement and are being followed. This may require the right to audit the processes of the outsource supplier or an attestation from the supplier that validates compliance. An attestation may be provided by the external auditors of the supplier or an independent reviewer.

Other issues related to outsourcing include declaring the jurisdiction of the agreement and which courts would hear any dispute related to the terms and conditions of the contract.

Some of the concerns that should be considered in relation to the risk of using an outsource supplier are:
• Hiring and training practices of the supplier
• Reporting and liaison between the outsourcing organization and supplier
• Time to respond to any incidents
• Liability for noncompliance with terms of the contract
• Nondisclosure of data or business practices
• Responding to requests from law enforcement
• Length of contract and terms for dissolution/termination of contract
• Location of data storage including backup data
• Separation between data and management of data of competing firms

When an organization is contracted to provide or deliver services or equipment, the risk of noncompliance with the agreement must be met through review, monitoring and enforcement of the contract terms. Any failure to meet contract terms must be identified and addressed as quickly as possible. This especially applies to the delivery of equipment that is not configured according to contractual agreements or may not provide the functionality promised during the contract negotiations and in the statement of work (SOW).

3.10.4 Data Management

Data are one of the most valuable assets of the organization; therefore, they must be protected at all times, in all forms and in all locations. Data must be protected in storage, transit, during processing and when displayed through access controls, encryption, backups and integrity checks, so that the information cannot be disclosed, altered or deleted in an unauthorized manner.

Data must be protected:
- In paper form (e.g., reports, scratch pads)
- On magnetic media (e.g., hard drives, universal serial bus [USB] storage)
- On optical drives
- In audio and video form
- On a screen
- When discarded

Protection of data during processing starts as soon as data are received. Input data should not be trusted and should be subject to validation checks before acceptance or processing. Input data validation should include:
- Range checks (e.g., allowable data values)
- Format checks (e.g., configuration of date)
- Special character checks (e.g., prevent script commands)
- Size (e.g., prevent buffer overflows [too much data] or incomplete data [too little data])
- Allowable values (e.g., months of the year)

Data validation may be done through a whitelist (a list of what data are allowed) or a blacklist (a list of what data are not allowed). In most cases, a whitelist is the preferred approach because it will only allow certain values, and most input data validation is based on static or infrequently changing values. The use of a common library for whitelisting can ensure that data are validated in a consistent manner by all applications that call the whitelist from the library. A blacklist has the disadvantage of requiring frequent updates and facing the challenge of canonicalization or the many different ways that input data can be formatted to avoid the blacklist rules.

Note: Canonicalization refers to the multiple ways to express the same term on a computer system. For example the expression *www.isaca.org* can also be written as:
- *http:/www.isaca.org*
- *http:/www.isaca.org/*
- *http://isaca.org*
- *https://www.isaca.org/Pages/default.aspx*
- *http://www%2eisaca%2eorg*

All of those entries would be processed the same way by the system. This is a challenge for blacklists because there are many ways to input the same command, and a blacklist specifies what input is not valid, requiring it to recognize all the ways the same input could be submitted. A failure to implement these rules correctly can lead to attacks such as directory traversal or bypassing firewall rules.

Processing data are further protected by ensuring that changes cannot be made that would affect the integrity, precision or accuracy of the data and data processing operations. This can be enforced by:
- Data checks and balances of input compared to output
- Checks of normal compared to abnormal levels of processing
- Anti-malware detection
- Segregation of duties
- A required process for transaction approval

One of the most important elements of data protection is control over the permissions and authorization levels of users that can access data and applications. Ensuring that users' permissions only include the least or lowest level of privilege to perform their job functions and only for the required time periods is key to protecting data from improper alteration or disclosure. This requires regular review of the access permissions of users and the revoking of permissions when a user leaves the organization or changes job roles.

Protecting data in storage includes actions such as isolation and the use of encryption. Isolation is used to keep sensitive data in separate networks or on systems that are not accessible to unauthorized personnel. This can be accomplished through network segmentation (including the use of firewalls and virtual local area networks [VLANs]), role-based access control, physical access controls and data encryption.

Cryptography

Encryption is used to protect the confidentiality of data by making data unreadable to unauthorized personnel, or protecting the integrity through hashing and digital signatures. Data encryption provides several risk management benefits, including:
• Confidentiality
• Integrity
• Proof of origin (nonrepudiation)
• Access control
• Authentication

Encryption alters the data from a plaintext (also known as cleartext), easily readable format into ciphertext of the data—a format that is unreadable without knowledge of the key used to protect (encrypt) the message.

Symmetric Algorithms

An example of symmetric key cryptography is shown in **exhibit 3.6**.

Symmetric key cryptographic systems are based on a symmetric encryption algorithm, which uses a secret key to encrypt the plaintext to the ciphertext and the same key to decrypt the ciphertext to the corresponding plaintext. In this case, the key is said to be symmetric because the encryption key is the same as the decryption key.

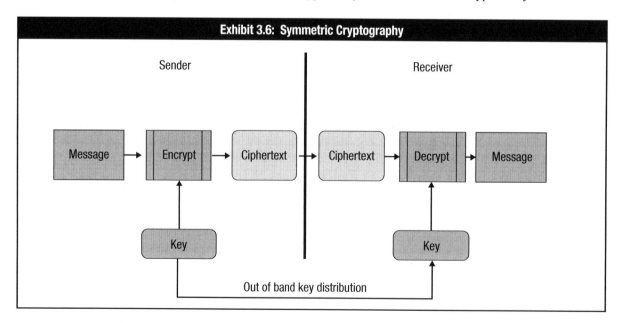

The most common symmetric key cryptographic system is the Data Encryption Standard (DES). DES is based on a public algorithm that operates on plaintext in blocks (strings or groups) of bits. This type of algorithm is known as a block cipher. DES uses blocks of 64 bits. A key of 56 bits is used for the encryption and decryption of plaintext. An additional 8 bits are used for parity checking. Any 56-bit number can be used as a key, and there are 72,057,594,037,927,936 possible keys in the key space.

DES is no longer considered a strong cryptographic solution because its entire key space can be brute-forced (every possible key tried) by large computer systems within a relatively short period of time. In this regard, private key cryptographic spaces of symmetric keys are susceptible to compromise. DES is being replaced with Advanced Encryption Standard (AES), a public algorithm that supports keys from 128 bits to 256 bits in size.

There are two main advantages to symmetric key cryptosystems such as DES or AES. The first is that the user has to remember/know only one key for both encryption and decryption. The second is that symmetric key cryptosystems are generally less complicated and, therefore, use up less processing power than asymmetric techniques. This makes symmetric key cryptosystems ideally suited for bulk data encryption. The major disadvantage of this approach is how to deliver the keys to those with whom you want to exchange data, particularly in e-commerce environments where customers are unknown, untrusted entities. Also, a symmetric key cannot be used to sign electronic documents or messages because the mechanism is based on a shared secret.

One form of advanced encryption algorithm is known as Triple DES, or DES3. Triple DES provides a relatively simple method of increasing the key size of DES to protect information without the need to design a completely new block cipher algorithm.

Asymmetric Algorithms
Asymmetric key cryptographic systems are a relatively recent development. In the 1970s, Diffie-Hellman developed a method of encryption based on two different keys. The two keys are mathematically related. One key is known as the private key while the other (which is based on the value of the private key) is known as the public key. Even if a person knows the public key, it is computationally unfeasible to determine the value of the private key. Therefore, the public key may be freely distributed without risking the compromise of the private key. The use of a public key allows many people to communicate securely with the holder of the corresponding private key, and for the holder of the private key to send a signed, authentic message to everyone that has the corresponding public key. Because asymmetric algorithms use a public and private key pair, they are commonly referred to as public key algorithms. Because each party using public key cryptography only needs one pair of keys (the private and the corresponding public key) the use of public key cryptography overcomes the weakness of scalability of symmetric key cryptography.

The main use of most asymmetric algorithms is to support the implementation of symmetric key algorithms and for digital signatures, as seen in **exhibit 3.7**.

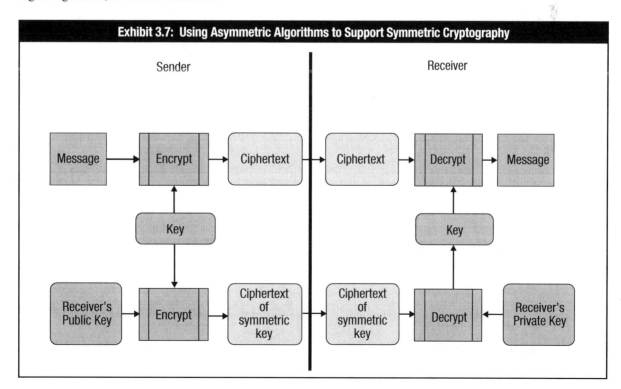

Exhibit 3.7: Using Asymmetric Algorithms to Support Symmetric Cryptography

A disadvantage to using asymmetric algorithms is they are very computationally intensive and very slow. They should not be used to encrypt large messages. Asymmetric cryptography should be used to encrypt short messages, such as a symmetric key, or to encrypt a hash of a message as a digital signature.

Message Integrity and Hashing Algorithms

The earliest networks used by computers to communicate were based on older voice-grade telephone cable and were of poor quality, were analog and had limited bandwidth. This meant that error correcting was needed to ensure data integrity and was originally performed using methods such as parity bits, checksums and cyclic redundancy checks (CRC). These methods were effective in detecting errors introduced by noise on the transmission line, but were not effective in preventing a malicious individual from intercepting and altering both the message and the integrity value. In that case, the recipient would not realize that the message had actually been altered en route.

Hashing algorithms are an accurate integrity check tool. The hash detects changes of even a single bit in a message. A hash algorithm will calculate a hash value (also known as a digest, fingerprint or thumbprint) from the entire input message. The output digest itself is a fixed length, so even though the input message can be of variable length, the output is always the same length. The length depends on the hash algorithm used. For example, MD5 generates a digest length of 128 bits; SHA1, a digest of 160 bits; and SHA512, a digest of 512 bits.

When a sender wants to send a message and ensure that it has not been affected by noise or network problems, the sender can compute the digest of the message and send it along with the message to the receiver. When the receiver receives the message and its digest, he/she computes the digest of the received message and ensures that the digest he/she computes is the same as the digest sent with the message. This is shown in **exhibit 3.8**.

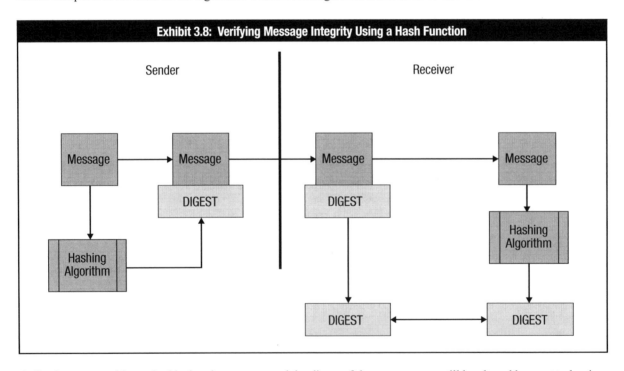

A disadvantage to this method is that the message and the digest of the message can still be altered by an attacker in the middle of the transmission path. Therefore, the solution is to combine two functions to protect both the message digest and to provide proof of origin of the message.

Digital Signatures

A digital signature combines a hash function with asymmetric encryption. The hash function provides proof of integrity, whereas the process of encrypting the hash of the message with a private key protects the hash from alteration and provides proof of origin of the message.

As shown in **exhibit 3.9**, the message itself is not confidential. Using this method, the receiver knows who the message came from because he/she was able to unlock the digital signature with the sender's public key, and therefore, it must have been encrypted with the sender's private key. Message integrity is protected by verifying that the hash of the received message is the same as the hash (digest) that was signed by the sender.

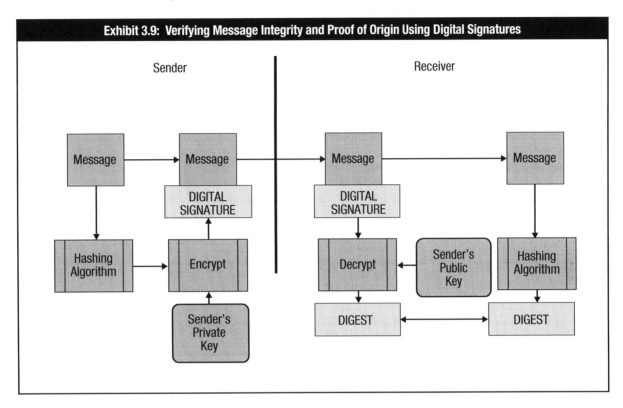

Exhibit 3.9: Verifying Message Integrity and Proof of Origin Using Digital Signatures

Certificates

The purpose of a certificate is to link a public key with its owner. A certificate is usually generated by a trusted third party known as a certificate authority (CA). The CA generates a certificate on behalf of the owner of a public key that the owner can use to prove that this public key belongs to them. If a sender sends a recipient a certificate created and signed by a CA, then the recipient can be confident that the public key in the certificate belongs to the sender. When the recipient opens a digital signature with that public key, he/she knows that the message was signed and sent by the sender.

The format of a certificate is based on the X.509 standard. This ensures that certificates can be accessed by most browsers and systems and that the certificates are of a common format even if issued by different CAs.

A certificate is valid for a defined period of time (often one year). However, if at any time during the period of time that the certificate is valid the owner wants to cancel the certificate, the certificate owner can notify the CA, and the CA will put the certificate on a certificate revocation list (CRL) that contains the list of certificates that are no longer valid but not yet expired.

Public Key Infrastructure

Public key infrastructure (PKI) refers to the implementation of asymmetric key cryptography. PKI implementation is based on a CA that manages all the certificates for the members of an implementation group. Any member of the group can gain access to the certificates of the other group members. This ensures that communication is secure and can only be accessed by the appropriate parties. A PKI may use an external CA, or a company may act as its own CA and manage the certificates for staff. When members of a PKI belong to different CAs, a cross-certification agreement should be in place in order for the members of one CA to be able to recognize the certificates issued by the other CA.

Summary of the Core Concepts of Cryptography

Cryptography has many moving parts and is a complex series of operations. In many cases, a combination of several of these features is used to accomplish various tasks.

- Symmetric encryption algorithms are excellent at providing confidentiality of large messages, but are difficult to use with key management. Therefore, symmetric algorithms are used to encrypt the message to be sent.
- Asymmetric algorithms are very slow, but are excellent for confidentially sending small messages. Therefore, asymmetric algorithms are used to send the symmetric key that was used to encrypt the message.
- Hashing algorithms are useful at ensuring message integrity. Therefore, a hash of the message is used to ensure the message is not changed en route.
- Asymmetric algorithms are useful at proving proof of origin. Therefore, they are used to sign the hash of the message, thereby creating a digital signature that can be used to prove both message integrity and proof of origin.
- Certificates provide assurance of the owner of a public key. Therefore, they are used to prove the authenticity of the owner and the correct public key of the owner or web site that is communicating.

3.10.5 Project and Program Management

The risk response to project and program management is based on addressing the risk that a project may not meet its objectives or that a failure of one or more projects may affect the delivery of a program.

Common risk responses to project risk include:
- Implementing a change control board to prevent scope creep
- Providing additional resources
- Prioritizing critical project tasks
- Reorganizing project resources/staffing
- Alteration of deliverables/time lines
- Replacement of project management
- Project cancellation
- Restarting the project
- Replacement of suppliers
- Renegotiating contracts

It is important to address project risk as soon as possible; otherwise, the project may continue to go off track and become increasingly difficult to correct.

3.10.6 The System Development Life Cycle

Risk must be evaluated and monitored at all points during the SDLC. This ensures that a system is designed, developed, tested, implemented and operated with adequate controls and protection. As the system moves through the phases of the SDLC, new risk may emerge and previous risk assessments may need to be revised. This requires the risk practitioner to work with the project team to identify and address risk and determine possible risk response. This is especially important when changes are proposed to an existing application, the risk practitioner must evaluate the potential impact of the change on the risk and security profiles of the application.

The risk practitioner should be alert to circumstances where the development team is not following the standards and policies of the organization regarding system development or the implementation of controls. This may represent a risk to the organization that should be escalated to management for resolution or consideration.

3.10.7 Business Continuity and Disaster Recovery Management

The business impact analysis (BIA) is a valuable tool that will be used by both the BCP/disaster recovery plan (DRP) team and the risk practitioner. The primary objective of the BIA is to identify risk and the impact of an incident, so that steps can be taken to prevent or respond effectively to an incident. Each incident is a learning experience, and the risk practitioner should review each incident to determine what lessons can be learned from the experience of the incident and what improvements can be made to preventive and detective controls and to incident handling procedures.

When a change is made to the risk environment, this may impact the BCP or DRP. The BCP/DRP team should be notified of any proposed changes in time to review and revise plans.

Risk management plans may have originally authorized a risk to be accepted, but that risk may now require remediation. Originally, a risk may have been addressed through BCPs, but the risk may now require new or modified controls. This will also affect the accuracy of the BCP.

3.10.8 IT Operations Management and Acquisition

IT-related initiatives are usually implemented as a project. When changes are made to a system, network or business process, the risk practitioner should be concerned with the preservation of the control infrastructure. After an IT system has been implemented, it must continue to be operated in a secure manner. This requires the hardening of systems being implemented and disabling default vendor accounts, enabling security functionality, and turning on the logging and audit functions of the system. Risk management should take care that the agreed secure standards are maintained.

IT departments are often the first to be notified of a system problem or potential breach, and they should be alert to system problems or abnormal conditions. The IT department should have clear procedures in place to respond to, and escalate, an incident.

Reports generated by monitoring should be provided to the risk owners to alert them to any potential problem or emerging trend.

The IT department is also responsible for backing up systems to enable system recovery. If the backups are stored offsite, it may be advisable to encrypt the backups, but provisions must be taken to ensure the encryption key is available when required.

> **Regular monitoring of controls should be performed to ensure that the controls are effectively mitigating risk.**

Backups should include:
- Files
- Transactions
- Operating systems (OSs)
- Databases
- Patches
- Configurations
- Access control lists
- Reports
- Applications

When purchasing IT equipment, IT should ensure that the equipment is delivered according to specifications, that default settings have been disabled, that security features have been enabled and that the equipment is installed according to the systems architecture and design.

3.10.9 Information Systems Architecture

There are different risk response options for each of the components of an information system. Each layer or device is best protected by using the solution and risk mitigation option that is best suited for that layer or device. For example, risk associated with a network layer attack are best mitigated using risk mitigation solutions best suited for network-based attacks. However, network-based solutions are not effective at mitigating an application-layer attack, so an application-layer vulnerability should be mitigated through application-based controls.

Platforms and Operating Systems

Recently, there have been several examples of hardware that has been infected with back doors and security vulnerabilities during the manufacturing or delivery process. These breaches have been found on network devices, point-of-sale terminals, applications and smartphones from numerous countries and vendors. Such vulnerabilities are not easy to detect. The risk practitioner should be aware of the risk of purchasing infected equipment and use trusted vendors or suppliers whenever possible. Purchasing equipment that has been tested and evaluated by an external entity using an internationally approved process, such as the Common Criteria (ISO 15408), may also provide a higher level of confidence that the equipment is secure. Once installed, the IT team must validate that maintenance hooks or back doors that vendors can use to gain access to, and monitor, the system have been secured or eliminated. All vendor-supplied default accounts and passwords should also be changed. When performing administration of hardware devices, a secure channel should be used and a form of strong authentication required.

No commercially available OS is absolutely secure; therefore, regardless of the underlying platform, it is necessary to implement strict controls over changes, patches and configurations. The systems should be hardened to disable all unnecessary services, and default accounts and passwords must be changed.

The culture of the IT operations team is also an important consideration. If the IT team is overconfident, they may not be careful enough to protect their systems and might leave them vulnerable. For example, if the UNIX support team in an organization is overconfident and does not put adequate controls into place, an attacker may be able to circumvent these weak controls and breach the security of this system. And, through that breach, they might also breach the Windows system, even though the Windows support team was very careful and had implemented better controls.

Securing platforms requires a patch management process that identifies, tests and schedules the implementation of the patches. Once a patch has been implemented, it should be monitored to ensure that all related systems, utilities and applications are still working correctly.

Applications

Application risk is often due to flaws or bugs in the coding of the application. This is especially true for web applications. Countermeasures to web application vulnerabilities are related to proper design, coding and testing. Resources, such as the Open Web Application Security Project *(www.owasp.org)*, are available that list common web application vulnerabilities and provide direction on how to mitigate and test for such vulnerabilities.

Applications should be designed to protect information using solutions such as:
• Masking to hide sensitive data
• Menus to restrict actions that are available to the user
• Drop-down boxes to limit input options
• Range checks to ensure expected or correct data values are submitted
• Balancing to ensure proper processing of transactions
• Logs to record all activity
• Access controls to restrict user access
• Certificates to authenticate entities
• Encryption to transmit sensitive data
• Documentation to facilitate maintenance
• Coding standards to support best practice in coding

Legacy applications may have other vulnerabilities due to older standards in coding or design. These may need to be protected through the use of middleware to isolate direct access and manage data input/output, network isolation and secure communications channels.

A common problem with applications is the process of error handling. If error messages are too verbose, they might provide information to an attacker that can be used to modify the attack. Errors should be coded and provide as little information as possible to the user.

When users attempt to log in to a network, application or system, they should be required to provide a user ID and password. The use of two-factor authentication is usually desirable, especially for critical systems, and after a few invalid login attempts, the user should be denied entry. This can be accomplished by locking the user ID, disconnecting the network connection or blocking the source IP address. User IDs may remain locked for a set period of time, or they may require resetting by an administrator.

Databases

Database security is essential because databases contain critical and sensitive information that is required to support business operations.

Database security is provided through:
• Encryption of sensitive data in the database
• Use of database views to restrict information available to a user
• Secure protocols to communicate with the database

- Content-based access controls that restrict access to sensitive records
- Restricting administrator-level access
- Efficient indexing to enhance data retrieval
- Backups of databases (shadowing, mirroring)
- Backups of transaction journals (remote journaling)
- Referential integrity
- Entity integrity
- Validation of input
- Defined data fields (schema)

Networks

Network security protects the integrity of data being transmitted to ensure that the data received are the same as the data sent; protects the confidentiality of data to ensure that confidential data were not disclosed to unauthorized parties during transmission; and protects availability to ensure that network communications were available for use whenever required by the organization.

Integrity of network communications is enabled through parity bits, checksums, hashing and digital signatures. Integrity checks can be performed by the network devices themselves, but in practice this is only performed at the end points because the networks are usually reliable. Integrity solutions include using authentication header (AH) mode of IP security (IPsec) and digital signatures.

Confidentiality of network data is accomplished through encryption. In some cases, the data may be encrypted by an application before being sent over a network. Often, it is responsibility of the network to encrypt the data. This can be done at the transport layer using Secure Socket Layer (SSL), Transport Layer Security (TLS) or Secure Shell (SSH). Confidentiality at the network layer is provided through the use of encapsulating security payload (ESP) mode of IPsec. Encryption at the datalink layer is provided through link encryption products such as Wi-Fi Protected Access II (WPA2) or serial encryption of microwave or other links between adjacent devices.

Network availability is provided through alternate routing, redundancy of cable and network devices, and load balancing. These solutions use multiple paths for data to travel and will still permit communications even in the event of a partial network failure.

The architecture of the network is important to isolate network segments and limit access to areas of the network and devices attached to the network. This is especially important when the network is used for devices such as point-of-sale terminals or other systems that handle sensitive data. Such devices should operate on a separate network segment to reduce the risk of data capture or modification. Another example of devices that should be on separate networks is supervisory control and data acquisition (SCADA) systems that are used to monitor and control remote devices (such as power plants, physical/environmental controls and pipelines). Network segmentation is often provided through the use of firewalls and gateways to control the flow of traffic between network segments. Networks also use layered defense to provide adequate protection by using several devices in serial to provide a series of hurdles that an attacker would have to overcome to successfully exploit the network. Devices used for layered defense can include intrusion detection and intrusion prevention systems (IDSs/IPSs), VLANs and bastion hosts.

The Internet is not secure, and any data sent over the Internet are subject to compromise unless they are protected using a virtual private network (VPN) or other solution.

In order for an organization to provide access to customers to products, data or services, the organization should design a demilitarized zone (DMZ). A DMZ is not fully secure and should only provide the minimal levels of access required. No sensitive data should be stored in a DMZ, and all devices in the DMZ should be hardened with limited functionality.

An extranet is a private network that resides on the Internet and allows a company to securely share business information with customers, suppliers or other businesses as well as to execute electronic transactions. This can be a secured closed network from a network provider, or a virtual network using the Internet, secured with certificates, two-factor authentication, or other methods.

It is good practice to use separate Internet connections for different functions: access from Internet to the web site, connection with trusted partners and external access for internal staff (e.g., web browsing, email). This allows tailored security measures for every function.

Administration of network devices should only be performed by authorized personnel and only through a change control and quality assurance process. Even a small error by a network administrator can lead to serious consequences. When performing remote administration using a protocol such as Simple Network Management Protocol (SNMP) versions 1 or 2, the connection should be over a secure encrypted connection (such as layer 2 tunneling protocol running over IPsec) because those protocols are not secure.

3.11 SYSTEMS CONTROL DESIGN AND IMPLEMENTATION

Systems should be designed to be secure, and the development process should ensure that security controls are built into the system and tested prior to deployment. As each part of a new system is developed or acquired, it should be tested. The risk of releasing a new or modified program that fails to meet business requirements or that results in a security breach is a significant factor in the risk management of the organization. Testing is the final opportunity the organization has to prevent a failure related to a poorly written program or improperly designed application. The objective of testing is to uncover any flaws or risk that may be hidden in the functionality or design of the application or system. The earlier a problem is found, the less expensive, and usually the more effective, the risk response will be.

3.11.1 Unit Testing

Unit testing is the testing of each individual component or piece of a system. This is the most basic level of test and is the best way to find a problem within the piece of code or piece of equipment being tested. As the components are integrated together into the larger system, it can be harder to find problems within an individual component.

Source codes for an application are tested by the developer using white box testing, meaning the developer has full access and visibility to the code itself. The developer can watch how the transaction flows through the code and verify that each line of code operates correctly.

When a device or an executable is purchased from a vendor, the tester cannot see into the code module or application/product to see how it operates. This is called black box testing. This type of testing monitors the behavior or function of the application/product, but does not show how it is actually delivering that function.

Code Review

Developers are required to follow the coding standards of the organization or good practices of the industry. All new or changed source code should be reviewed by a third party to validate compliance with standards and good coding practices. This process can detect unauthorized changes made by the programmer and detect inadequate error handing, input validation or documentation. The process of third-party review should be done by an impartial, knowledgeable and, if possible, anonymous party. The third-party review should be based on standards, not only programmer preferences.

3.11.2 Integration Testing/System Testing

After the individual components of a system have been tested, the entire system should be tested. This will show how the components work when they are integrated or joined together along with the interfaces between the components and the overall operation of the system.

Integration testing tests the system in relation to its overall environment. For example, it can show whether the new or changed system accepts data properly from upstream systems and integrates properly with downstream systems.

Unit testing and initial integration testing is often performed in a separate area from the final system testing. This separates the developer from the final testing and acceptance process and helps ensure that the components being tested are not being modified by the developer during the testing process.

A part of this process is to test the security functions designed into the project and ensure that the system has been built according to the design. And, if not built according to design, it should be ensured that deviations from the plan have been documented and approved. **Exhibit 3.10** describes different testing options.

Exhibit 3.10: Options for System Testing	
Test	**Description**
Recovery testing	Checks the system's ability to recover after a software or hardware failure
Security testing	Verifies that the modified/new system includes provisions for appropriate access controls and does not introduce any security holes that may compromise other systems
Stress/volume testing	Tests an application with large quantities of data to evaluate its performance during peak hours
Volume testing	Studies the impact on the application by testing with an incremental volume of records to determine the maximum volume of records (data) that the application can process
Stress testing	Studies the impact on the application by testing with an incremental number of concurrent users/services on the application to determine the maximum number of concurrent users/services the application can process
Performance testing	Compares the system's performance to other equivalent systems using well-defined benchmarks

3.11.3 Version Control

An important part of system development is version control. There is always the risk that a change to a system will overwrite or bypass functionality that was changed in an earlier version. It is also important to make sure that when multiple teams are working on new versions of a program, changes made by another team are not overlooked. Version control tracks the current version of the software and checks for previous changes when a new change is implemented. Version control can also ensure that the correct version of the source code of the software is being used in compiling the object code for production.

3.11.4 Regression Testing

Regression testing involves testing the changes to a program to discover any new problems in the operation of the program that were caused by the changes (i.e., a change in one part of the program caused a problem in another part of the program) or whether the changes made have inadvertently removed previous fixes.

3.11.5 Test Data

There is significant risk related to test data. Ideally, test data should never use sensitive production data. All sensitive data elements should be displayed in a manner that makes it unreadable (obfuscated), such as masking sensitive information by displaying it as special characters or altering the data itself. This will prevent the disclosure of sensitive data to unauthorized personnel during the testing process. Test data should be complete and allow the testing of all possible process functions and error handling.

3.11.6 Fuzzing

Fuzzing is the process of testing input fields (controls) of a program. The process of fuzzing will test allowable values, often the limit of the acceptable range of values and test values that are beyond the allowable values. This permits the tester to observe whether the input validation and process integrity controls are working correctly.

3.11.7 Separation of Development From Production

Good risk management separates the system development area from the production area. System programmers should not be working directly in production, and separation of the networks and physical areas used by developers can protect the organization from unauthorized or inadequately tested changes.

3.11.8 Quality Assurance

Prior to approval to implement a new or modified program, the program should be approved through a user acceptance test (UAT). This validates whether the program is meeting user requirements and expectations. A UAT can highlight problems with the functionality, training or process flow that may not have been detected earlier in the process.

The process of quality assurance validates that the program will deliver the expected functionality in a reliable and secure manner, and that the program has been developed and documented according to organizational standards and good practice. Quality assurance is a planned and systematic pattern of all actions necessary to provide adequate confidence that an item or product conforms to established technical requirements.

Part of the process to migrate a new or modified system or application into production is to lock down the code and ensure that the final version of the program cannot be modified inadvertently after it has been approved for implementation or during the final testing process. Some organizations call this final version the "gold code" or "locked code." The responsibility for the final testing and preparation for implementation is often conducted by business representatives. The business analysts and representatives may test a beta version of the program as a simulation of the operational environment and conditions. After the final tests of the code have been conducted, the program will be scheduled for implementation.

3.11.9 Fallback (Rollback)

When implementing a new system or making changes to an existing system, there is always the risk that the change will not work successfully. This could be due to technical problems, software problems or other unexpected challenges. The project team should have a fallback plan so that it is possible to roll back to the earlier program or configuration where possible. This will allow the project team to mitigate the risk if system deployment does not go as planned. To mitigate the risk of downtime for mission-critical systems, good practices dictate that the tools and applications required to reverse the migration are available prior to attempting the production cutover.

3.11.10 Challenges Related to Data Migration

When migrating from one system to another or modifying an existing system, it may be necessary to perform data conversion or migration. This is a risk to the integrity and availability of the data. The risk practitioner should assess the process used for data migration and advise of associated risk. Some considerations for data conversion are listed in **exhibit 3.11**.

Exhibit 3.11: Data Conversion Key Considerations	
Consideration	**Guidelines**
Completeness of data conversion	The total number of records from the source database is transferred to the new database (assuming the number of fields is the same).
Data integrity	The data are not altered manually, mechanically or electronically by a person, program or substitution or by overwriting in the new system. **Note:** Integrity problems also include errors due to transposition and transcription errors and problems transferring particular records, fields, files and libraries.
Storage and security of data under conversion	Data are backed up before conversion for future reference or any emergency that may arise out of data conversion program management. **Note:** An unauthorized copy or too many copies can lead to misuse, abuse or theft of data from the system.
Data consistency	The field/record called for from the new application should be consistent with that of the original application. **Note:** This enables consistency in repeatability of the testing exercise.
Business continuity	The new application should be able to continue with newer records as added (or appended) and help in ensuring seamless business continuity.

3.11.11 Changeover (Go-live) Techniques

There are several methods commonly used to enable the transition between versions of a system or application or from one system to another. These include:
• Parallel changeover
• Phased changeover
• Abrupt changeover

Parallel Changeover

A parallel changeover is done by running both systems simultaneously. This allows the project team to test the reliability and performance of the new system and ensure that it is working correctly before removing the old system from service. The benefits of a parallel changeover are that it minimizes the risk of a failed cutover to the new systems, allows testing of the performance of the new system without affecting production, and allows staff to train on and become familiar with the new system before it is in full production.

A disadvantage of a parallel changeover is the cost of maintaining both systems at the same time and the challenge of ensuring that the data are consistent between both systems.

Exhibit 3.12 demonstrates the principles of a parallel changeover.

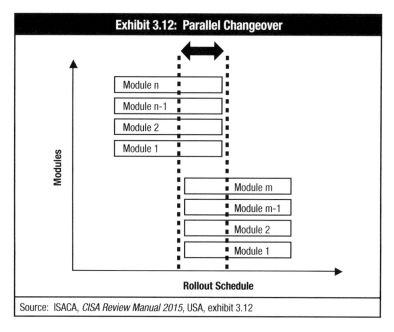

Exhibit 3.12: Parallel Changeover

Source: ISACA, *CISA Review Manual 2015*, USA, exhibit 3.12

Phased Changeover

A phased changeover is conducted by replacing individual components or modules of the old system with new or modified components. This reduces the risk by gradually rolling out the new modules without impacting the entire system. Depending on the type of system, however, this type of cutover is not always possible.

Some of the risk factors that may exist in the phased changeover include:
• Resource challenges on the IT side (to be able to maintain two unique environments such as hardware, OSs, databases and code) and on the operations side (to be able to maintain user guides, procedures and policies; definitions of system terms, etc.)
• Maintaining consistency of data on multiple systems or locations
• Extension of the project life cycle to cover two systems
• Change management for requirements and customizations to maintain ongoing support of the older system

Exhibit 3.13 shows the principles of a phased changeover.

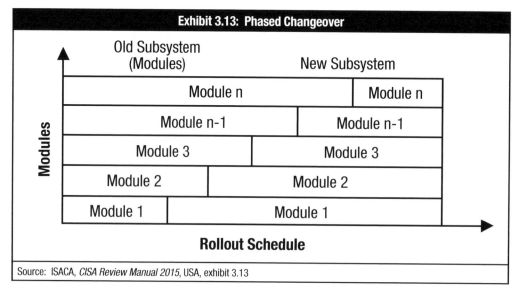

Exhibit 3.13: Phased Changeover

Source: ISACA, *CISA Review Manual 2015*, USA, exhibit 3.13

Abrupt Changeover

An abrupt changeover is the riskiest changeover process because it may be impossible to roll back if the project fails. In this process, the old system is replaced entirely with the new or modified system at once, and the operation of the old system is discontinued.

The steps in an abrupt changeover are:
1. Convert files and programs; perform test runs on the test bed.
2. Install new hardware, OS, application system and migrated data.
3. Train employees or users in groups.
4. Schedule operations and test runs for go-live or changeover.

Exhibit 3.14 demonstrates the principles of an abrupt changeover.

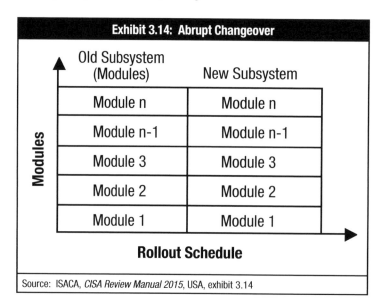

Exhibit 3.14: Abrupt Changeover

Source: ISACA, *CISA Review Manual 2015*, USA, exhibit 3.14

The potential risk associated with abrupt changeover includes:
• Asset safeguarding
• Data integrity
• System effectiveness
• System efficiency
• Change management challenges (depending on the configuration items considered)
• Duplicate or missing records (duplicate or erroneous records may exist if data cleansing is not done correctly)

3.11.12 Postimplementation Review
A postimplementation review should be conducted on all projects to derive lessons learned from the project and enable more effective results for future projects. The review is useful for determining whether the project was properly designed, developed, implemented and managed and that the appropriate controls have been built into the system.

A postimplementation review should include the following:
• Assess the adequacy of the system:
  – Does the system meet user requirements and business objectives?
  – Have controls been adequately defined and implemented?
• Evaluate the projected cost benefits or ROI measurements.
• Develop recommendations that address the system's inadequacies and deficiencies.
• Develop a plan for implementing the recommendations.
• Assess the development project process:
  – Were the chosen methodologies, standards and techniques followed?
  – Were appropriate project management techniques used?
  – Is the risk of operating the system within acceptable risk levels?
The project development team and appropriate end users perform a post-project review jointly after the project has been completed and the system has been in production for a sufficient time period to assess its effectiveness.

3.11.13 Project Closeout
Projects should have a finite life: At some point, the project is closed and the new or modified system is handed over to the users and/or system support staff. **Exhibit 3.15** shows the steps that should be taken when closing a project.

Step	Action
	Exhibit 3.15: Project Closeout Steps
1	Assign any outstanding issues to individuals responsible for remediation and identify the related budget, if applicable.
2	Assign custody of contracts, and archive or pass on documentation to those who will need it.
3	Survey the project team, development team, users and other stakeholders to: • Identify any lessons learned that can be applied to future projects. • Include content-related criteria such as: – Performance fulfillment and project-related incentives – Fulfillment of additional objectives – Adherence to the schedule and costs • Include process-related criteria such as: – Quality of the project teamwork – Relationships to relevant environments
4	Conduct reviews in a formal process, such as a post-project review, in which lessons learned and an assessment of project management processes used are documented and can be referenced, in the future, by other project managers or users working on projects of similar size and scope.
5	Complete a postimplementation review after the project has been in use (or in production) for some time—long enough to realize its business benefits and costs—and measure the project's overall success and impact on the business units.

NOTE: The project sponsor should be satisfied that the system produced is acceptable and ready for delivery.

3.12 IMPACT OF EMERGING TECHNOLOGIES ON DESIGN AND IMPLEMENTATION OF CONTROLS

The organization should encourage a proactive approach to the use of new technologies. This requires the risk practitioner to be alert to the development of new technologies and to assess the risk that those technologies may pose if they were to be implemented into the organization. The risk practitioner should consider the use of discovery scanners to detect unauthorized implementation of new devices or technologies that have not yet been reviewed and approved for use. Policies regarding the use or introduction of new technologies should be developed, and personnel should be aware of the processes required to submit a new technology for consideration for use.

3.13 CONTROL OWNERSHIP

There may be areas in which responsibility and ownership of risk is not defined or clearly stated. All risk should have an owner, and that owner must be at a management level that can make decisions on behalf of the organization. The risk practitioner should communicate with the risk owners to ensure that they are aware of the risk responses that have been implemented, the risk responses that are still pending implementation and the level of risk that the organization is currently facing. If the risk owner is not willing to accept the current level of risk, then the owner must mandate the implementation of new risk mitigation.

3.14 RISK MANAGEMENT PROCEDURES AND DOCUMENTATION

The risk practitioner should be interested in developing risk management solutions that effectively and efficiently manage risk at acceptable levels. This is often accomplished through the use of controls, and, as discussed previously, controls usually consist of a combination of administrative, technical and physical elements. The risk practitioner can play a key role in ensuring that the controls are set up and operating correctly and are being maintained, evaluated and reported to management on a regular basis.

Control management procedures include:
• The proper installation of the controls
• The creation of policies and procedures to support the operation of the controls
• The implementation of a change management procedure to ensure controls are configured correctly
• The training of staff to monitor, manage and review control operation
• Assignment of responsibility for control monitoring
• Assignment of responsibility for incident investigation
• Creation of a schedule for review and reporting on the control

The controls should be documented along with the justification for the control, the owner of the control and the reporting schedule.

As risk response projects are completed and milestones reached, the risk register should be updated to reflect the changes. Some risk entries may be closed to show a completed risk mitigation project, while other entries, such as a status update, will require modification and review. The risk register should also show the progress of testing and the attainment of milestones during the progress of a risk mitigation project. The risk register should always be an accurate and up-to-date review of risk and risk mitigation projects.

3.15 RISK RESPONSES AND THE RISK ACTION PLAN

As products and risk responses are implemented, the risk practitioner should ensure that the project was implemented according to the intent and design of the project architects, or if changes to the design were necessary, that the changes have been reviewed to ensure that they did not erode or diminish the effectiveness of the security control and that a suitable risk level has been achieved.

3.16 SUMMARY

The organization should be aware of its risk, know the assessed level of the risk and take steps to document its risk in the risk register. Based on the risk assessment, the risk response options are examined and a suitable risk response is selected. Projects and programs are developed to implement the risk response, and the results are tracked in the risk register. Controls are designed and implemented, and the level of risk is reduced to a level that the management of the organization can accept. Changes in risk and the effectiveness of controls should continue to be monitored, and this information should be reported to management.

ENDNOTES

[1] International Organization for Standardization (ISO)/International Electrotechnical Commission (IEC); *ISO/IEC 27001:2013 Information technology—Security techniques—Information security management systems—Requirements*, Switzerland, 2013

Page intentionally left blank

Idintifies the Risk Profile = RR

Chapter 4: Risk and Control Monitoring and Reporting

Section One: Overview

Section Two: Contents

ISACA. All Rights Reserved.

Section One: Overview

DOMAIN DEFINITION

Continuously monitor and report on IT risk and controls to relevant stakeholders to ensure the continued efficiency and effectiveness of the IT risk management strategy and its alignment to business objectives.

LEARNING OBJECTIVES

The objective of this domain is to ensure that the CRISC candidate has the knowledge necessary to:
• Differentiate between key risk indicators (KRIs) and key performance indicators (KPIs)
• Describe data extraction, aggregation and analysis tools and techniques
• Compare different monitoring tools and techniques
• Describe various testing and assessment tools and techniques

CRISC EXAM REFERENCE

This domain represents 22 percent of the CRISC exam (approximately 33 questions).

TASK AND KNOWLEDGE STATEMENTS

TASKS

There are seven tasks within this domain that a CRISC candidate must know how to perform. These relate to IT risk identification.

T4.1   Define and establish key risk indicators (KRIs) and thresholds based on available data, to enable monitoring of changes in risk.

T4.2   Monitor and analyze key risk indicators (KRIs) to identify changes or trends in the IT risk profile.

T4.3   Report on changes or trends related to the IT risk profile to assist management and relevant stakeholders in decision making.

T4.4   Facilitate the identification of metrics and key performance indicators (KPIs) to enable the measurement of control performance.

T4.5   Monitor and analyze key performance indicators (KPIs) to identify changes or trends related to the control environment and determine the efficiency and effectiveness of controls.

T4.6   Review the results of control assessments to determine the effectiveness of the control environment.

T4.7   Report on the performance of, changes to, or trends in the overall risk profile and control environment to relevant stakeholders to enable decision making.

KNOWLEDGE STATEMENTS

The CRISC candidate should be familiar with the task statements relevant to each domain in the CRISC job practice. The tasks are supported by 41 knowledge statements that delineate each of the areas in which the risk practitioner must have a good understanding in order to perform the tasks. Many knowledge statements support tasks that cross domains.

The CRISC candidate should have knowledge of:
1. Laws, regulations, standards and compliance requirements
2. Industry trends and emerging technologies
3. Enterprise systems architecture (e.g., platforms, networks, applications, databases and operating systems)
4. Business goals and objectives
5. Contractual requirements with customers and third-party service providers
6. Threats and vulnerabilities related to:
    6.1. Business processes and initiatives
    6.2. Third-party management
    6.3. Data management
    6.4. Hardware, software and appliances

6.5. The system development life cycle (SDLC)

6.6. Project and program management

6.7. Business continuity and disaster recovery management (DRM)

6.8. Management of IT operations

6.9. Emerging technologies

7. Methods to identify risk

8. Risk scenario development tools and techniques

9. Risk identification and classification standards, and frameworks

10. Risk events/incident concepts (e.g., contributing conditions, lessons learned, loss result)

11. Elements of a risk register

12. Risk appetite and tolerance

13. Risk analysis methodologies (quantitative and qualitative)

14. Organizational structures

15. Organizational culture, ethics and behavior

16. Organizational assets (e.g., people, technology, data, trademarks, intellectual property) and business processes, including enterprise risk management (ERM)

17. Organizational policies and standards

18. Business process review tools and techniques

19. Analysis techniques (e.g., root cause, gap, cost-benefit, return on investment [ROI])

20. Capability assessment models and improvement techniques and strategies

21. Data analysis, validation and aggregation techniques (e.g., trend analysis, modeling)

22. Data collection and extraction tools and techniques

23. Principles of risk and control ownership

24. Characteristics of inherent and residual risk

25. Exception management practices

26. Risk assessment standards, frameworks and techniques

27. Risk response options (i.e., accept, mitigate, avoid, transfer) and criteria for selection

28. Information security concepts and principles, including confidentiality, integrity and availability of information

29. Systems control design and implementation, including testing methodologies and practices

30. The impact of emerging technologies on design and implementation of controls

31. Requirements, principles, and practices for educating and training on risk and control activities

32. Key risk indicators (KRIs)

33. Risk monitoring standards and frameworks

34. Risk monitoring tools and techniques

35. Risk reporting tools and techniques

36. IT risk management best practices

37. Key performance indicator (KPIs)

38. Control types, standards, and frameworks

39. Control monitoring and reporting tools and techniques

40. Control assessment types (e.g., self-assessments, audits, vulnerability assessments, penetration tests, third-party assurance)

41. Control activities, objectives, practices and metrics related to:

41.1. Business processes

41.2. Information security, including technology certification and accreditation practices

41.3. Third-party management, including service delivery

41.4. Data management

41.5. The system development life cycle (SDLC)

41.6. Project and program management

41.7. Business continuity and disaster recovery management (DRM)

41.8. IT operations management

41.9. The information systems architecture (e.g., platforms, networks, applications, databases and operating systems)

SELF-ASSESSMENT QUESTIONS

4-1   The **MOST** important reason to maintain key risk indicators (KRIs) is because:

      A. complex metrics require fine-tuning.
      B. threats and vulnerabilities change over time.
      C. risk reports need to be timely.
      D. they help to avoid risk.

4-2   Which of the following choices is the **BEST** measure of the operational effectiveness of risk management process capabilities?

      A. Key performance indicators (KPIs)
      B. Key risk indicators (KRIs)
      C. Base practices
      D. Metric thresholds

4-3   During a data extraction process, the total number of transactions per year was forecasted by multiplying the monthly average by twelve. This is considered:

      A. a controls total.
      B. simplistic and ineffective.
      C. a duplicates test.
      D. a reasonableness test.

4-4   The **BEST** test for confirming the effectiveness of the system access management process is to map:

      A. access requests to user accounts.
      B. user accounts to access requests.
      C. user accounts to human resources (HR) records.
      D. the vendor database to user accounts.

4-5   Which of the following choices provides the **BEST** assurance that a firewall is configured in compliance with an enterprise's security policy?

      A. Review the actual procedures.
      B. Interview the firewall administrator.
      C. Review the parameter settings.
      D. Review the device's log file for recent attacks.

4-6   One way to verify control effectiveness is by determining:

      A. its reliability.
      B. whether it is preventive or detective.
      C. the capability of providing notification of failure.
      D. the test results of intended objectives.

4-7   An enterprise has implemented a tool that correlates information from multiple systems. This is an example of a monitoring tool that focuses on:

A. transaction data.
B. configuration settings.
C. system changes.
D. process integrity.

4-8   Which of the following methods is the **MOST** effective way to ensure that outsourced service providers comply with the enterprise's information security policy?

A. Periodic audits
B. Security awareness training
C. Penetration testing
D. Service level monitoring

ANSWERS TO SELF-ASSESSMENT QUESTIONS

Correct answers are shown in **bold**.

4-1 A. While most key risk indicator (KRI) metrics need to be optimized in respect to their sensitivity, the most important objective of KRI maintenance is to ensure that KRIs continue to effectively capture the changes in threats and vulnerabilities over time.

      B. Threats and vulnerabilities change over time, and KRI maintenance ensures that KRIs continue to effectively capture these changes.

      C. Risk reporting timeliness is a business requirement, but is not a driver for KRI maintenance.

      D. Risk avoidance is one possible risk response. Risk responses are based on KRI reporting.

4-2 **A. Key performance indicators (KPIs) are assessment indicators that support the judgment of the process performance of a specific process.**

      B. Key risk indicators (KRIs) only provide insights into potential risk that may exist or be realized within a concept or capability that they monitor.

      C. Base practices are activities that, when consistently performed, contribute to achieving a specific process purpose. However, base practices need to be complemented by work products to provide reliable evidence about the performance of a specific process.

      D. Metric thresholds are decision or action points that are enacted when a KPI or KRI reports a specific value or set of values.

4-3 A. The described test does not ensure that all transactions have been extracted.

      B. While simplistic, the reasonableness test is a valid foundation for more elaborate data validation tests.

      C. The described test does not identify duplicate transactions.

      D. Reasonableness tests make certain assumptions about the information as the basis for more elaborate data validation tests.

4-4 A. Mapping access requests to user accounts confirms that all access requests have been processed; however, the test does not consider user accounts that have been established without the supporting access request.

      B. Mapping user accounts to access requests confirms that all existing accounts have been approved.

      C. Mapping user accounts to human resources (HR) records confirms whether user accounts are uniquely tied to employees.

      D. Mapping vendor records to user accounts may confirm valid accounts on an e-commerce application, but is flawed because it does not consider user accounts that have been established without the supporting access request.

4-5 A. While procedures may provide a good understanding of how the firewall is supposed to be managed, they do not reliably confirm that the firewall configuration complies with the enterprise's security policy.

      B. While interviewing the firewall administrator may provide a good process overview, it does not reliably confirm that the firewall configuration complies with the enterprise's security policy.

      C. A review of the parameter settings provides a good basis for comparison of the actual configuration to the security policy and reliable audit evidence documentation.

      D. While reviewing the device's log file for recent attacks may provide indirect evidence about the fact that logging is enabled, it does not reliably confirm that the firewall configuration complies with the enterprise's security policy.

4-6   A.   Reliability is not an indication of control strength; weak controls can be highly reliable, even if they do not meet the control objective.

        B.   The type of control (preventive or detective) does not help determine control effectiveness.

        C.   Notification of failure does not determine control strength.

        D.   Control effectiveness requires a process to verify that the control process worked as intended and meets the intended control objectives.

4-7   **A.   Monitoring tools focusing on transaction data generally correlate information from one system to another, such as employee data from the human resources (HR) system with spending information from the expense system or the payroll system.**

        B.   Configuration settings are generally compared against predefined values and not based on the correlation between systems.

        C.   System changes are compared from a previous state to the current state.

        D.   Process integrity is confirmed within the system.

4-8   **A.   Regular audits can identify gaps in information security compliance.**

        B.   Training can increase user awareness of the information security policy, but is not more effective than auditing.

        C.   Penetration testing can identify security vulnerability, but cannot ensure information compliance.

        D.   Service level monitoring can only identify operational issues in the enterprise's operational environment.

NOTE: For more self-assessment questions, you may also want to obtain a copy of the *CRISC™ Review Questions, Answers & Explanations Manual 2015*, which consists of 400 multiple-choice study questions, answers and explanations, and the *CRISC™ Review Questions, Answers & Explanations Manual 2015 Supplement*, which consists of 100 new multiple-choice study questions, answers and explanations.

SUGGESTED RESOURCES FOR FURTHER STUDY

In addition to the resources cited throughout this manual, the following resources are suggested for further study in this domain (publications in **bold** are stocked in the ISACA Bookstore):

Beasley, Mark S.; Bruce C. Branson; Bonnie V. Hancock; *Developing Key Risk Indicators to Strengthen Enterprise Risk Management*, Committee of Sponsoring Organizations of the Treadway Commission (COSO), USA, 2010

ISACA, *COBIT® 5 for Risk*, **USA, 2013,** *www.isaca.org/cobit*

ISACA, *COBIT® Self-assessment Guide: Using COBIT® 5*, **USA, 2013,** *www.isaca.org/cobit*

ISACA, *The Risk IT Practitioner Guide*, **USA, 2009**

Section Two: Content

4.0 OVERVIEW

The risk response is designed and implemented based on a risk assessment that was conducted at a single point in time. Risk changes; controls can become less effective, the operational environment may change, and new threats, technologies and vulnerabilities may emerge. Because of the changing nature of risk and associated controls, ongoing monitoring is an essential step of the risk management life cycle as seen in **exhibit 4.1**.

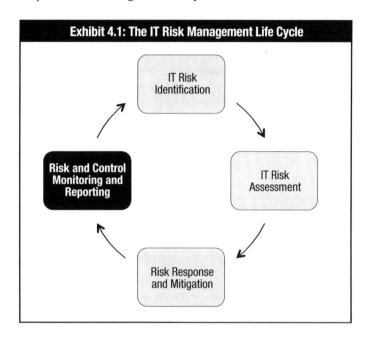

The effectiveness of the risk response and selected controls must be measureable. This provides management with information that shows the effectiveness of the IT risk management effort and justifies the expenses of supporting the controls. Management is also required to prove due care and due diligence in protecting the assets of the organization and meeting regulatory requirements.

Management must also be aware of changes in risk over time and the impact of changes in the risk environment. Risk should be a part of strategic planning so that changes in risk can be anticipated and mitigated even as new strategic plans are unveiled, designed and implemented.

The trends or ongoing changes in risk and control effectiveness need to be documented and tracked so that an unhealthy change in risk can be noticed before it becomes a severe issue.

The use of key risk indicators (KRIs) and key performance indicators (KPIs) can help management to identify and track the maturity and effectiveness of the IT risk management program over time. With regular reporting, management will be able to identify and monitor a situation that requires attention in a timely manner.

4.1 KEY RISK INDICATORS

Risk indicators are used to measure levels or risk in comparison to defined risk thresholds and alert the organization when a risk level approaches a high or unacceptable level of risk. The purpose of a risk indicator is to set in place tracking and reporting mechanisms that alert staff to a developing or potential risk.

A KRI is a subset of risk indicators that are highly relevant and possess a high probability of predicting or indicating important risk. Examples of KRIs include:
• Quantity of unauthorized equipment or software detected in scans
• Number of instances of service level agreements (SLAs) exceeding thresholds
• High average downtime due to operational incidents
• Average time to deploy new security patches to servers
• Excessive average time to research and remediate operations incidents
• Number of desktops/laptops that do not have current antivirus signatures or have not run a full scan within scheduled periods

KRIs support the following aspects of risk management:[1]
• Risk appetite, by validating the organization's risk appetite and risk tolerance levels
• Risk identification, by providing an objective means for identifying risk
• Risk mitigation, by providing a trigger for investigating an event or providing corrective action
• Risk culture, by helping the organization focus on important, relevant areas
• Risk measurement and reporting, by providing objective and quantitative risk information
• Regulatory compliance, by providing data that can be used as an input for operation risk capital calculations

4.1.1 KRI Selection
A common mistake when implementing KRIs—other than selecting too many KRIs—includes choosing KRIs that:
• Are not linked to specific risk
• Are incomplete or inaccurate due to unclear specifications
• Are difficult to measure
• Are difficult to aggregate, compare and interpret
• Provide results that cannot be compared over time
• Are not linked to goals

When selecting KPIs, good metrics are SMART:
• **Specific**—Based on a clearly understood goal; clear and concise
• **Measureable**—Able to be measured; quantifiable (objective), not subjective
• **Attainable**—Realistic; based on important goals and values
• **Relevant**—Directly related to a specific activity or goal
• **Timely**—Grounded in a specific time frame

The selection of the appropriate set of KRIs benefits the organization by:
• Providing an early warning (forward-looking) signal that a high risk is emerging to enable management to take proactive action (before the risk actually becomes a loss)
• Providing a backward-looking view on risk events that have occurred, enabling risk responses and management to be improved
• Enabling the documentation and analysis of trends
• Providing an indication of the enterprise's risk appetite and tolerance through metric setting (i.e., KRI thresholds)
• Increasing the likelihood of achieving the enterprise's strategic objectives
• Assisting in continually optimizing the risk governance and management environment

When working with management to determine appropriate KRIs, the risk practitioner should work with all relevant stakeholders to ensure greater buy-in and ownership. Risk indicators should be identified for all stakeholders and should not focus solely on IT. IT-based metrics should be aligned as much as possible with other metrics used in the organization to report to stakeholders.

Other factors that can influence the selection of KRIs include:
• **Balance**: Risk indicators should be balanced and cover:
  – Lag indicators (indicating risk after events have occurred)
  – Lead indicators (indicating which controls are in place to prevent events from occurring)
  – Trends (analyzing indicators over time or correlating indicators to gain insights)
• **Root cause**: Selected indicators should drill down to the root cause of events, not just the symptoms

4.1.2 KRI Effectiveness

KRI effectiveness takes into consideration the following criteria:
- **Impact**: Indicators of risk with high business impact are more likely to be KRIs.
- **Effort**: For different indicators that are equivalent in sensitivity, the one that is easier to measure and maintain is preferred.
- **Reliability**: The indicator must possess a high correlation with the risk and be a good predictor or outcome measure.
- **Sensitivity**: The indicator must be representative of risk and capable of accurately indicating risk variances.
- **Repeatable**: A KRI must be repeatable and able to be measured on a regular basis to show trends and patterns in activity and results.

4.1.3 KRI Optimization

To ensure accurate and meaningful reporting, KRIs need to be optimized to ensure that: (1) the correct data are being collected and reported on, and (2) that the KRI thresholds are set correctly. KRIs that are reporting on data points that cannot be controlled by the enterprise, or are not alerting management at the correct time to an adverse condition, must be adjusted (optimized) to be more precise, more relevant or more accurate.

Exhibit 4.2 describes examples in which KRIs may need to be optimized.

| Exhibit 4.2: Examples in Which KRIs Should Be Optimized ||
Metric Criterion	Description
Sensitivity	Management has implemented an automated tool to analyze and report on access control logs based on severity; the tool generates excessively large amounts of results. Management performs a risk assessment and decides to configure the monitoring tool to report only on alerts marked "critical."
Timing	Management has implemented strong segregation of duties (SoD) within the enterprise resource planning (ERP) system. One monitoring process tracks system transactions that violate the defined SoD rules before month-end processing is completed so that suspicious transactions can be investigated before reconciliation reports are generated.
Frequency	Management has implemented a key control that is performed multiple times a day. Based on a risk assessment, management decides that the monitoring activity can be performed weekly because this will capture a control failure in sufficient time for remediation.
Corrective action	Automated monitoring of controls is especially conducive to being integrated into the remediation process. This can often be achieved by using existing problem management tools, which help prioritize existing gaps, assign problem owners and track remediation efforts.

4.1.4 KRI Maintenance

Because the organization's internal and external environments are constantly changing, the risk environment is also highly dynamic, and the set of KRIs needs to be changed over time. Each KRI is related to the risk appetite and tolerance levels of the enterprise. KRI trigger levels should be defined at a point that enables stakeholders to take appropriate action in a timely manner.

4.2 KEY PERFORMANCE INDICATORS

A KPI is a measure that determines how well the process is performing in enabling the goal to be reached.

A KPI is an indicator of whether a goal will likely be reached and a good indicator of capabilities, practices and skills. It measures an activity goal, which is an action that the process owner must take to achieve effective process performance.

For example, a KPI may indicate that an error rate of five percent is acceptable. An error rate higher than five percent would be unacceptable and require escalation and some form of response.

KPIs are used to set benchmarks for risk management goals and to monitor whether those goals are being attained. Management sets its risk management goals according to its risk acceptance level and desired cost-benefit analysis.

A KPI should be:
• Valuable to the business
• Tied to a business function or service
• Under the control of management
• Quantitatively measured
• Used repeatedly in different reporting periods

Examples of KPIs include:
• Network availability
• Customer satisfaction
• Number of complaints resolved on first contact
• Response time for data
• Number of employees that attended awareness sessions

A KPI places an emphasis on a certain process that should be a good indicator of the health of the overall process. For example, measuring the time required to deploy security patches may be a good indicator of the effectiveness of the risk management process related to IT operations and serve as a good competitive benchmark. This is directly controllable by management because management can set policies; manage the change control process; mandate objectives; and control staff resourcing, technology, staff training and attention to customer focus. It is also quantitative and easily measured on a monthly or semiannual basis, and therefore, the results can be compared over time. Another example of a KPI related to risk management is the frequency at which risk assessments are being conducted. The KPI could verify whether systems, projects, training programs, etc., are being assessed for effectiveness according to policy and mandated frequency.

A KPI is often used on a chart or graph to track compliance and to report to management in a clear, easily understood manner.

4.2.1 Using KPIs With KRIs

KPIs and KRIs are often used in conjunction with one another to measure performance and mitigate risk. However, they should be distinguished from one another. KPIs help to identify underperforming aspects of the organizations and areas of the business that may require additional resources and attention. KRIs provide early warnings of increased risk within the organization.[2] While somewhat different, the two can be used together in risk management.

For example, the risk associated with unpatched systems may lead to a serious breach of customer data, availability and financial loss. Therefore, the organization develops a policy to apply all critical patches within 30 days. The development of this policy is an example of a KPI. On a scheduled basis, the organization tracks and reports to senior management on whether that KPI has been met and reports on the average time to deploy critical security patches. However, to detect a trend or developing problem, the organization may also set a KRI that will represent a threshold or level of performance that is getting dangerously close to exceeding the KPI, but has not yet done so. In this example, the KRI may be set at 25-day patch deployments.

If an organization tracks the time between the release of a patch and its deployment within the organization on a monthly basis, then a report that shows the average time taken could be developed, as shown in **exhibit 4.3**.

Through this chart, a manager can see that the time to deploy patches on a UNIX system is increasing each month and is in danger of exceeding the 30-day threshold. By setting a KRI of 25 days, management can see that the time to deploy a UNIX patch has now increased to the point of triggering an alert. This alert should precipitate a review of the UNIX patching procedures, and management should investigate the situation to enable preventive action and address the slow deployment process before the deployments of these patches violates the policy.

This example shows how KRIs and KPIs can be used to monitor and track the attainment of goals. When measuring the success of a risk management program, KPIs must be based on meaningful criteria that can help management track the overall success of the controls put in place to support risk response.

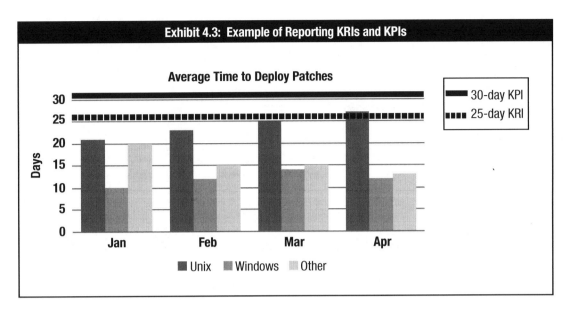

Exhibit 4.3: Example of Reporting KRIs and KPIs

Average Time to Deploy Patches

4.3 DATA COLLECTION AND EXTRACTION TOOLS AND TECHNIQUES

The risk practitioner may use various sources of data to monitor and report on risk.

Internal data sources include:
• Audit reports
• Incident reports
• User feedback
• Observation
• Interviews with management
• Security reports
• Logs

4.3.1 Logs

Logs are one of the most popular ways to capture and store data for analysis. Logs may contain sensitive information, and they may be needed for forensic purposes. Logs should:
• Be configured in a way that prevents alteration or deletion, and access should be limited to authorized personnel only
• Be configured to capture and retain the most pertinent information for an adequate period of time regarding the actions of a person or process that performed the activity
• Try to capture event data close to the source of an event so that it is easier to associate the activities of a process or individual with the recorded event

If a log contains too many data from too many disparate sources, it may be difficult to notice significant individual events.

Analysis of log data can identify security violations and be instrumental in forensics investigations, but it can also alert the organization to a developing attack or multiple attempts to break in. This may be used to identify the source of an attack and assist in strengthening controls where necessary. Time synchronization of log entries can assist with correlation of events from multiple sources.

Analysis of log data and control activity should answer the following questions:
• Are the controls operating correctly?
• Is the level of risk acceptable?
• Are the risk strategy and controls aligned with business strategy and priorities?
• Are the controls flexible enough to meet changing threats?
• Is the correct risk data being provided in a timely manner?
• Is the risk management effort aiding in reaching corporate objectives?
• Is the awareness of risk and compliance embedded into user behaviors?

4.3.2 Security Event and Incident Management

Security event and incident management (SEIM) systems are data correlation tools that capture data from multiple sources and provide data analysis and reports to management on system, application and network activity and possible security events. These tools can be used to detect compliance violations, identify known attacks and provide reporting.

As shown in **exhibit 4.4**, a SEIM system gathers data from multiple sources and correlates and analyzes that data to develop reports for management on the wider picture of security across the systems of the organization and to highlight relationships among activity on various parts of the network or systems. The reports are generated based on the types of incidents observed, the timing of incidents, the chronological sequence of events and the source of the activity. When an event is logged on only one system, it can be difficult to notice that the problem may actually be related to a larger issue. A SEIM system can help identify that. Therefore, the use of a SEIM system can give management a more accurate overview of the risk profile of the organization as a whole.

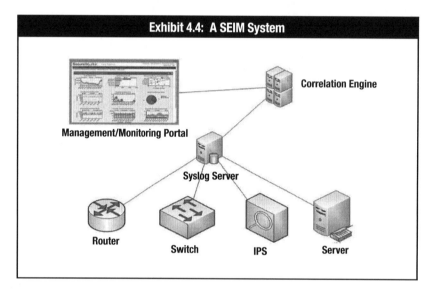

Exhibit 4.4: A SEIM System

4.3.3 Integrated Test Facilities

One method of monitoring the performance and operation of an application is to use an integrated test facility (ITF). An ITF is a testing methodology that processes test data through production systems. It is used to test whether the systems are operating correctly, whether a potential problem exists with the processes and to detect a risk condition. An organization can set up several fictitious customers or transactions that are processed along with real data. This allows business analysts to observe the operation of the production systems and ensure correct processing.

4.3.4 External Sources of Information

External data sources can include:
- Media reports
- Computer emergency response team (CERT) advisories
- Security company reports
- Regulatory bodies
- Peer organizations

There are a variety of tools that a risk practitioner can use when monitoring risk. These include free reports from antivirus and security companies. Other reports, such as the Verizon Data Breach Investigations report,[3] are issued annually to provide a review of risk and exposure factors and may assist in developing and presenting a risk management solution to management. Other sources of information on incidents and risk come from government sources such as local or national computer incident response teams (CIRT) and nonprofit organizations.

4.4 CHANGES TO THE IT RISK PROFILE

Risk management is a never-ending cycle that recognizes the dynamic nature of risk and the need to continuously monitor and assess risk. When the organization is in the risk and control monitoring phase, it often finds that, due to the changes in the risk environment, it has to start again with risk identification, assessment and response.

Organizations need to pursue the development and maturing of their risk management process and carefully monitor, report on and improve the risk profile and culture of the organization. Risk management, like health and safety and information security, is a responsibility of everyone in the organization.

The risk profile is based on the overall risk posture of the organization and is reflected by its attentiveness to monitoring the effectiveness of risk mitigating controls, how proactive it is in identifying and preventing risk, and its approach to developing a risk culture.

The risk profile measures the effectiveness of the risk management program, including measuring compliance with laws and policies, reporting on the status of risk mitigation projects, and identifying and addressing emerging threats.

Changes in the risk profile are the result of:
• New technologies
• Changes to business procedures
• Mergers or acquisitions
• New or revised regulations
• Changes in customer expectations
• Actions of competitors
• Effectiveness of risk awareness programs

4.5 MONITORING CONTROLS

According to the *COBIT 5 for Risk* management practice MEA01.01 *Establish a monitoring approach*, the process of setting up an IS control monitoring process requires the risk practitioner to "engage with stakeholders to establish and maintain a monitoring approach to define the objectives, scope and method for measuring business solution and service delivery and contribution to enterprise objectives. Integrate this approach with the corporate performance management system." The risk monitoring and evaluation approach should be integrated with the overall corporate performance management systems to ensure alignment between IT risk and business risk. This would prevent an acceptance of risk within IT that exceeds the risk acceptance criteria set by the business that could otherwise have an unacceptable or unidentified level of impact on business mission or operations.

The controls mandated through risk management must align with IT security and related policies of the enterprise. The IS control monitoring function ensures that IT security requirements are being met; standards are being followed; and staff is complying with the policies, practices and procedures of the organization.

The steps to monitoring controls are:
1. Identify and confirm risk control owners and stakeholders.
2. Engage with stakeholders and communicate the risk and information security requirements and objectives for monitoring and reporting.
3. Align and continually maintain the information security monitoring and evaluation approach with the IT and enterprise approaches.
4. Establish the information security monitoring process and procedure.
5. Agree on a life cycle management and change control process for information security monitoring and reporting.
6. Request, prioritize and allocate resources for monitoring information security.

The risk practitioner must remember that the purpose of a control is to mitigate a risk. Therefore, the purpose of control monitoring is to verify whether the control is effectively addressing the risk—not only to see whether the control itself is working. For example, a firewall protects one network from another by regulating network traffic. The assessment of the firewall is not only to see whether the firewall is operating correctly and managing traffic, but also to see whether the network is being adequately protected and the risk mitigated.

> **The monitoring and metrics of managing a control must be based on data relevant to the risk and the overall performance of the device.**

The purpose of risk monitoring and evaluation is to collect, validate and evaluate business, IT and process goals and metrics; to monitor that processes are performing against agreed-on performance and conformance goals and metrics; and provide reporting that is systematic and timely. This requires the gathering of data related to risk management from various sources in a timely and accurate manner. After the data have been validated for integrity, an analysis of the data against specific performance targets can be performed. Properly done, this will provide a succinct, all-around view of IT performance within the enterprise monitoring system.

When the results of the monitoring indicate an area of noncompliance or unacceptable performance, the risk practitioner should recommend the use of mitigating activities, such as implementation of new controls, adjustment or enforcement of existing controls, or changes to a business process to address the risk environment.

The risk practitioner should have the goal to continuously monitor, benchmark and improve the IT control environment and control framework to meet organizational objectives. This can be met through timely reporting and follow-up, and continuous monitoring and evaluation of the control environment. A process of continuous monitoring and reporting will enable management to identify control deficiencies and inefficiencies and to initiate improvement actions.

The monitoring of controls and the risk management framework may be done through self-assessment or independent assurance reviews. When risk action plans are required to address a risk, the plan should be monitored to ensure appropriate risk management practices in alignment with risk appetite and tolerance. Assurance activities performed by independent entities should be independent from the function, group or organization being monitored. These entities should have the necessary skills and competence to perform assurance and adhere to codes of ethics and professional standards.

Risk monitoring and control activities should be in place, and any exceptions should be reported promptly, followed up on and analyzed, and corrective actions should be prioritized and implemented. As the risk environment changes and internal control systems are affected by changes in business and IT risk, any gap in the risk environment must be evaluated and changes made where necessary.

Sources of control monitoring information include:
- Network operations centers (NOC)
- Command and control centers
- Continuous control monitoring
- Periodic testing
- Independent assessments

The risk practitioner should encourage management and each process owner to take positive ownership of control improvement through a continuing program of self-assessment to evaluate the completeness and effectiveness of management's control over processes, policies and contracts.

4.6 CONTROL ASSESSMENT TYPES

The effectiveness of the control monitoring process depends on the accuracy and completeness of the data provided for control evaluation. The risk practitioner must ensure that the data they receive are genuine and free from errors or misstatements. Data that can be retrieved directly by the risk practitioner are preferable to data that have been provided by a third party.

The risk practitioner should encourage local ownership of risk and control monitoring. Local ownership creates a risk culture in which local managers accept responsibility for risk and monitoring the behaviors of their staff and enables a faster detection of violations and security incidents.

4.6.1 IS Audits

Audit teams provide independent and objective review of the effectiveness and appropriateness of the control environment. Information provided by IS auditors can underline the need for control enhancement and bring risk to the attention of management. By working with the audit teams, the risk practitioner can align the risk management program with audit and will often provide supporting data to the IS auditors. Recommendations provided by IS audit will often require the attention of the risk practitioner through the updating of risk action plans, the risk register and further control enhancement.

4.6.2 Vulnerability Assessments and Penetration Tests

Risk monitoring includes regular vulnerability assessments and penetration tests that can be conducted either internally or externally.

Vulnerability Assessments

A vulnerability assessment is a valuable tool used to identify any gaps or misconfigurations in the security profile of the organization. It is a methodical review of security to ensure that the systems are hardened and that there are no unnecessary open ports or services available that could be used as an attack vector by an adversary or misused by an internal employee.

A vulnerability assessment is often a thorough review of an entire system or facility, and it is intended that the assessment will provide a good, complete review of all security controls, including both technical and nontechnical controls. For example, a review of a firewall includes the review of the configuration (technical control), the change control process for managing the configuration (nontechnical control), the process for monitoring the firewall logs, the system architecture to ensure the firewall cannot be bypassed, and the training and competence levels of the firewall administrators. Through such a vulnerability assessment, management can gain a solid report on the strength and effectiveness of the risk management program.

There are many open source and commercial tools that can be used to perform a vulnerability assessment, and many web sites that list known vulnerabilities with common applications, operating systems (OSs) and utilities.

The problem with a vulnerability assessment can be the number of false positives or noise that it generates. This can make it difficult to determine the actual severity of the problems. Therefore, a penetration test is necessary.

Penetration Tests

A penetration test is a targeted attempt to break into a system or application (or in a physical test, to break into a building). Using the results of a vulnerability assessment, the tester selects a potential vulnerability and tries to exploit that vulnerability. If the penetration tester is able to break in, then the vulnerability is real and must be mitigated. If the tester is unable to break in, then there is a good chance that the vulnerability is not serious and does not require mitigation.

A penetration tester often uses the same tools used by hackers to try to break into systems. This means that the results are quite real and provide meaningful results. However, this also poses a risk to the organization because these tools can be hazardous to use and may result in system failure or compromise. Therefore, it is of utmost importance that such tests are only conducted with management approval and through the use of a defined methodology and oversight.

> The results of the penetration test should be provided to management for follow-up and review.

4.6.3 Third-party Assurance

The use of a third party to provide assurance or attestation of the validity of the information security program of the organization can be a valuable tool to earn customer and shareholder confidence. This attestation may be based on an external IS audit or a certification of compliance with an internationally recognized standard, such as COBIT 5 or ISO/IEC 27001, or an industry standard, such as PCI DSS. The third party is responsible for evaluating the processes of the subject organization and validating compliance with the requirements of the standard.

Another source for attestation is based on the Statement on Standards for Attestation Engagements Number 16 (SSAE 16). SSAE 16 was developed by the American Institute of Certified Public Accountants (AICPA) to replace the older SAS 70 standard. SSAE 16 complies with the international service organization reporting standard International Standards for Assurance Engagements Number 3402 (ISAE 3402). This form of third-party assurance is used by many organizations when relying on cloud or third-party service suppliers.

4.7 RESULTS OF CONTROL ASSESSMENTS

The risk practitioner is expected to provide a report to management on the status of the risk management program and the overall risk profile of the organization. This requires the review of the effectiveness of, and compliance with, the controls in the organization. Controls may need adjustment, replacement or removal depending on the changes in the risk environment and the acceptance and appropriateness of the controls.

The effectiveness of control monitoring is dependent on the:
• Timeliness of the reporting—Are data received in time to take corrective action?
• Skill of the data analyst—Does the analyst have the skills to properly evaluate the controls?
• Quality of monitoring data available—Are the monitoring data accurate and complete?
• Quantity of data to be analyzed—Can the risk practitioner find the important data in the midst of all the other log data available?

4.7.1 Maturity Model Assessment and Improvement Techniques

The risk practitioner must be committed to continuous improvement of the risk management program. A mature and healthy risk management program will be better at preventing, detecting and responding to security events and risk scenarios. Maturity and growth comes from practice and attention to learning from past events. As the risk practitioner develops the skills, tools and team necessary for better risk management, the consistency in how risk is identified, assessed, mitigated and monitored will also mature and develop.

Routine use of a capability maturity model (CMM) shows the maturation of the risk management process year over year. A CMM starts with level zero—undefined and *ad hoc* activities and progresses—through the steps of defining and following a program; learning and enhancement of the program; and finally, a mature program that represents stable, quality processes and reliable, accurate information.

Risk Profile living Documentation of what type of change Could Cause Riskpro New Technologies Change to Business procedure mergers or acquisition

An example of the CMM is shown in **exhibit 4.5.**

Exhibit 4.5: Example of the Capability Maturity Model		
COBIT 5 ISO/IEC 15504-based Capability Levels	**Meaning of the COBIT 5 ISO/IEC 15504-based Capability Levels**	**Context**
5 Optimized	The previously described predictable process is continuously improved to meet relevant current and projected business goals.	Enterprise view/Corporate knowledge
4 Predictable	The previously described established process now operates within defined limits to achieve its process outcomes.	Enterprise view/Corporate knowledge
3 Established	The previously described managed process is now implemented using a defined process that is capable of achieving its process outcomes.	Enterprise view/Corporate knowledge
2 Managed	The previously described performed process is now implemented in a managed fashion (planned, monitored and adjusted) and its work products are appropriately established, controlled and maintained.	Instance view/Individual knowledge
1 Performed	The implemented process achieves its process purpose.	Instance view/Individual knowledge
0 Incomplete	The process is not implemented or fails to achieve its process purpose. At this level, there is little or no evidence of any systematic achievement of the process purpose.	Instance view/Individual knowledge
Adapted from ISACA, COBIT 5, USA, 2012, figure 20		

4.8 CHANGES TO THE IT RISK PROFILE

To evaluate the risk control effectiveness and efficiency and determine the IT risk profile of the organization, the following areas should be continually monitored:

- New assets
- Changes to the scope of risk assessment
- Changes in business priorities, assets, services, products and operations
- Risk acceptance levels
- New threats (internal and external)
- Newly discovered vulnerabilities
- Possibility of increased risk due to aggregation of threats and vulnerabilities
- Incidents
- Logs and other data sources
- People and the morale of the organization
- Changes in the supply chain (e.g., mergers, political turmoil, transportation problems, natural events, bankruptcies of vendors)
- Changes in the financial markets

Other areas that should be monitored by the risk practitioner include:
- Regulations and legal changes
- Actions of competitors (benchmarking and standards)
- Total cost of ownership (TCO) of assets
- Impact from external events
- Availability of staff/resources

It is important to review the IT risk management objectives and goals on a periodic (annual) basis to ensure that the goals of risk management continue to be aligned with the goals and objectives of senior management. This includes review of the criteria used for monitoring (are the correct things being monitored and logged?), the thresholds used for KPIs and KRIs (are they still reflective of management levels of concern?), the policies and strategies of risk (are they still aligned with business strategy?), the reporting schedule, and the key stakeholders.

The results of the past year should demonstrate maturing and improvement in the IT risk profile of the organization, including the completion of risk response and mitigation activities, the training of staff, the success of awareness programs, improved response time to incidents, timely rollout of patches, and better alignment and communication among management, audit, business continuity, physical security and information security departments.

Risk is owned by management, but the risk practitioner has a key role in ensuring that management is aware of the current IT risk profile of the organization and that risk is being managed in a way that meets management objectives. Throughout this phase of the IT risk management process, the risk practitioner works with the risk owners, IT, third parties, incident response teams and auditors to monitor risk, and from that, evaluate the effectiveness and efficiency of the control framework. As incidents occur, lessons learned are used to improve the risk management process. These lessons may pertain to better knowledge, staffing, technical controls, procedures, monitoring and response programs. All of these benefits can help avoid future problems, enable the impact of future incidents to be minimized and sustain business operations.

4.9 SUMMARY

Proper and effective management of risk is essential to protecting the assets and attaining the goals of the organization. Throughout this review manual, the principles of risk management have been discussed and practical steps have been used to demonstrate the benefits, objectives and concepts associated with the identification, assessment, reporting, response and monitoring of risk. The objective of risk management is to discover and address all risk in an appropriate manner and ensure that the organization has reduced risk to acceptable levels and does not become a victim of a risk that should have been identified and mitigated.

As risk is identified and assessed, the risk owners select the appropriate response to the risk and create risk action plans to implement or modify controls selected to mitigate risk. As controls to mitigate risk are designed and developed, the risk owner also mandates the development of the ability to monitor and report on the effectiveness of the controls. Without the ability to monitor the controls, there is no assurance that the controls are working correctly, or that they are effectively mitigating the risk.

Regular monitoring and reporting on risk is essential to management, and the use of KPIs and KRIs assists management in the monitoring of trends, compliance and issues related to risk. The focus continues to turn toward more continuous monitoring of risk by every manager and every department. This will aid in timely reporting and response to risk events and facilitate the development of more effective and efficient risk management.

Risk management is a never-ending process. As the external and internal environments change, as technology changes, and as the nature of attacks and attackers evolves, so also does the need to revisit the risk management effort, and reassess risk, revise risk response, and improve the risk culture and awareness of risk throughout the organization.

IT risk and controls should be continuously monitored and reported on to relevant stakeholders to ensure the continued efficiency and effectiveness of the IT risk management strategy and its alignment to business objectives.

ENDNOTES

[1] Fraser, John; Betty Simkins; *Enterprise Risk Management: Today's Leading Research & Best Practices for Tomorrow's Executives*, John Wiley & Sons, USA, 2010

[2] Beasley, Mark S.; Bruce C. Branson; Bonnie V. Hancock; *Developing Key Risk Indicators to Strengthen Enterprise Risk Management*, Committee of Sponsoring Organizations of the Treadway Commission (COSO), USA, 2010.

[3] Verizon; *2014 Data Breach Investigations Report*, USA, 2014, *www.verizonenterprise.com/DBIR/2014*

GENERAL INFORMATION

The CRISC certification is designed to meet the growing demand for professionals who can integrate enterprise risk management (ERM) with discrete IS control skills. The technical skills and practices the CRISC certification promotes and evaluates are the building blocks of success in this growing field, and the CRISC designation demonstrates proficiency in this role.

REQUIREMENTS FOR CERTIFICATION

To earn the CRISC designation, the following requirements must be met:
1. Pass the CRISC exam.
2. Submit an application (within five years of the passing date) with verified evidence of a minimum of at least three years of cumulative work experience performing the tasks of a CRISC professional across at least two CRISC domains. Of the two required domains, one must be risk-related, either Domain 1 (IT Risk Identification) or 2 (IT Risk Assessment). There will be no substitutions or experience waivers. A processing fee of US $50 must accompany all applications.
3. Adhere to the ISACA Code of Professional Ethics.
4. Agree to comply with the CRISC continuing education policy.

Please note that certification application decisions are not final as there is an appeal process for certification application denials. Appeals undertaken by a certification exam taker, certification applicant or by a certified individual are undertaken at the discretion and cost of the exam taker, applicant or individual. Inquiries regarding denials of certification can be sent to *certification@isaca.org*.

SUCCESSFUL COMPLETION OF THE CRISC EXAM

The exam is open to all individuals who wish to take it. Successful exam candidates are not certified until they apply for certification (and demonstrate that they have met all requirements) and receive approval from ISACA.

EXPERIENCE IN RISK AND INFORMATION SYSTEMS CONTROL

Work experience must be gained within the 10-year period preceding the application for certification or within five years from the date of initially passing the exam. An application for certification must be submitted within five years from the passing date of the CRISC exam. All experience must be verified independently with employers.

> **Note:** A CRISC candidate may choose to take the CRISC exam prior to meeting the experience requirements.

DESCRIPTION OF THE EXAM

The CRISC Certification Committee oversees the development of the exam and ensures the currency of its content. The exam consists of 150 multiple-choice questions that cover the CRISC job practice domains. The job practice was developed and validated using prominent industry leaders, subject matter experts and industry practitioners.

REGISTRATION FOR THE CRISC EXAM

The CRISC exam will be administered in June and December in 2015. Please refer to the *ISACA Exam Candidate Information Guide* at *www.isaca.org/examguide* for specific registration dates, deadlines and registration forms, as well as important key information for exam day. Exam registrations can be placed online at *www.isaca.org/examreg*.

CRISC PROGRAM ACCREDITATION UNDER ISO/IEC 17024:2003

The American National Standards Institute (ANSI) has voted to continue the accreditation for the CISA, CISM, CGEIT and CRISC certifications under ISO/IEC 17024:2003, General Requirements for Bodies Operating Certification Systems of Persons. ANSI, a private, nonprofit organization, accredits other organizations to serve as third-party product, system and personnel certifiers.

ISO/IEC 17024 specifies the requirements to be followed by organizations certifying individuals against specific requirements. ANSI describes ISO/IEC 17024 as "expected to play a prominent role in facilitating global standardization of the certification community, increasing mobility among countries, enhancing public safety, and protecting consumers."

ANSI's accreditation:
• Promotes the unique qualifications and expertise ISACA's certifications provide
• Protects the integrity of the certifications and provides legal defensibility
• Enhances consumer and public confidence in the certifications and the people who hold them
• Facilitates mobility across borders or industries

Accreditation by ANSI signifies that ISACA's procedures meet ANSI's essential requirements for openness, balance, consensus and due process. With this accreditation, ISACA anticipates that significant opportunities for CISAs, CISMs, CGEITs and CRISCs will continue to open in the United States and around the world.

PREPARING FOR THE CRISC EXAM

The CRISC exam evaluates a candidate's practical knowledge of the job practice domains listed in this manual and online at *www.isaca.org/criscjobpractice.* That is, the exam is designed to test a candidate's knowledge and experience of the proper application of IT risk and IS control best practices. Since the exam covers a broad spectrum of risk and IS control issues, candidates are cautioned not to assume that reading CRISC study guides and reference publications will fully prepare them for the exam. CRISC candidates are encouraged to refer to their own experiences when studying for the exam and refer to CRISC study guides and reference publications for further explanation of concepts or practices with which the candidate is not familiar.

No representation or warranties are made by ISACA in regard to CRISC exam study guides, other ISACA publications, references or courses assuring candidates' passage of the exam.

TYPES OF EXAM QUESTIONS

CRISC exam questions are developed with the intent of measuring and testing practical knowledge and the application of general concepts and standards. All questions are multiple choice and are designed for one best answer.

Every question has a stem (question) and four options (answer choices). The candidate is asked to choose the correct or best answer from the options. The stem may be in the form of a question or incomplete statement. In some instances, a scenario may also be included. These questions normally include a description of a situation and require the candidate to answer two or more questions based on the information provided. The candidate is cautioned to read each question carefully. An exam question may require the candidate to choose the appropriate answer based on a qualifier, such as **MOST** important or **BEST**. In every case, the candidate is required to read the question carefully, eliminate known incorrect answers and then make the best choice possible. To gain a better understanding of the types of question that might appear on the exam and how these questions are developed, refer to the Item Writing Guide available at *www.isaca.org/itemwriter.*

ADMINISTRATION OF THE EXAM

ISACA has contracted with an internationally recognized testing agency. This not-for-profit corporation engages in the development and administration of credentialing exams for certification and licensing purposes. It assists ISACA in the construction, administration and scoring of the CRISC exam.

SITTING FOR THE EXAM

Candidates are to report to the testing site at the time indicated on their admission ticket. NO CANDIDATE WILL BE ADMITTED TO THE TEST CENTER ONCE THE CHIEF EXAMINER BEGINS READING THE ORAL INSTRUCTIONS. Candidates who arrive after the oral instructions have begun will not be allowed to sit for the exam and will forfeit their registration fee. To ensure that candidates arrive in time for the exam, it is recommended that candidates become familiar with the exact location of, and the best travel route to, the exam site prior to the date of the exam. Candidates can use their admission tickets only at the designated test center on the admission ticket.

To be admitted into the test site, candidates must bring the email printout OR a printout of the downloaded admission ticket and an acceptable form of photo identification such as a driver's license, passport or government ID. This ID must be a current and original government-issued identification that is not handwritten and that contains both the candidate's name as it appears on the admission ticket and the candidate's photograph. Candidates who do not provide an acceptable form of identification will not be allowed to sit for the exam and will forfeit their registration fee. Candidates are not to write on the admission ticket.

The following conventions should be observed when completing the exam:
• Do not bring study materials (including notes, paper, books or study guides) or scratch paper or notepads into the exam site. For further details regarding what personal belongings can (and cannot) be brought into the test site, please visit *www.isaca.org/criscbelongings.*
• Candidates are not allowed to bring any type of communication, surveillance or recording device (including, but not limited to, cell phones, tablets, smart glasses, smart watches, mobile devices, etc.) into the test center. If candidates are viewed with any such device during the exam administration, their exams will be voided and they will be asked to immediately leave the exam site.
• Candidates who leave the testing area without authorization or accompaniment by a test proctor will not be allowed to return to the testing room and will be subject to disqualification.
• Candidates should bring several no. 2 pencils since pencils will not be provided at the exam site.
• As exam venues vary, every attempt will be made to make the climate control comfortable at each exam venue. Candidates may want to dress to their own comfort level.
• Read the provided instructions carefully before attempting to answer questions. Skipping over these directions or reading them too quickly could result in missing important information and possibly losing credit points.
• Mark the appropriate area when indicating responses on the answer sheet. When correcting a previously answered question, fully erase a wrong answer before writing in the new one.
• Remember to answer all questions since there is no penalty for wrong answers. Grading is based solely on the number of questions answered correctly. Do not leave any question blank.
• Identify key words or phrases in the question (**MOST, BEST, FIRST** …) before selecting and recording the answer.
• The chief examiner or designate at each test center will read aloud the instructions for entering information on the answer sheet. It is imperative that candidates include their exam identification number as it appears on their admission ticket and any other requested information on their exam answer sheet. Failure to do so may result in a delay or errors.

BUDGETING TIME
The following are time-management tips for the exam:
• It is recommended that candidates become familiar with the exact location of, and the best travel route to, the exam site prior to the date of the exam.
• Candidates should arrive at the exam testing site at the time indicated on their admission ticket. This will give the candidate time to be seated and get acclimated.
• The exam is administered over a four-hour period. This allows for a little over 1.5 minutes per question. Therefore, it is advisable that candidates pace themselves to complete the entire exam. In order to do so, candidates should complete an average of 37.5 questions per hour.
• Candidates are urged to record their answers on their answer sheet. No additional time will be allowed after the exam time has elapsed to transfer or record answers should candidates mark their answers in the question booklet. The exam will be scored based on the answer sheet recordings only.

RULES AND PROCEDURES
• Candidates are asked to sign the answer sheet to protect the security of the exam and maintain the validity of the scores.
• Candidates who are discovered engaging in any kind of misconduct—such as giving or receiving help; using notes, papers or other aids; attempting to take the exam for someone else; using any type of communication, surveillance or recording device, including cell phones, during the exam administration; removing test materials, answer sheet or notes from the testing room; or attempting to share test questions or answers or other information contained in the exam (as such are the confidential information of ISACA)—will have their exam voided and be asked to leave the exam site. Candidates who leave the testing area without authorization or accompaniment by a test proctor will not be allowed to return to the testing room and will be subject to disqualification. Candidates who continue to write the exam after the proctor signals the end of the examination time may have their examination voided. Candidates may not access items stored in the personal belongings area until they have completed their exams. The testing agency will report all cases of misconduct to the respective ISACA Certification Committee for Committee review in order to render any decision necessary. Sharing the confidential test items subsequent to the exam will also be considered misconduct resulting in a voided examinations score.

• Candidates may not take the exam question booklet after completion of the exam.
• Candidates are not permitted to access items stored in the personal belongings area during the exam.

GRADING THE CRISC EXAM AND RECEIVING RESULTS

The exam consists of 150 items. Candidate scores are reported as a scaled score. A scaled score is a conversion of a candidate's raw score on an exam to a common scale. ISACA uses and reports scores on a common scale from 200 to 800. A candidate must receive a score of 450 or higher to pass the exam. A score of 450 represents a minimum consistent standard of knowledge as established by ISACA's CRISC Certification Committee. A candidate receiving a passing score may then apply for certification if all other requirements are met.

Passing the exam does not grant the CRISC designation. To become a CRISC, each candidate must complete all requirements, including submitting an application for certification.

The CRISC examination contains some questions which are included for research and analysis purposes only. These questions are not separately identified and the candidate's final score will be based only on the common scored questions.

A candidate receiving a score less than 450 is not successful and can retake the exam by registering and paying the appropriate exam fee for any future exam administration. To assist with future study, the result letter each candidate receives includes a score analysis by content area. There are no limits to the number of times a candidate can take the exam.

Approximately eight weeks after the test date, the official exam results will be mailed to candidates. Additionally, with the candidate's consent during the registration process, an email containing the candidates pass/fail status and score will be sent to paid candidates. This email notification will only be sent to the address listed in the candidate's profile at the time of the initial release of the results. To ensure the confidentiality of scores, exam results will not be reported by telephone or fax. To prevent the email notification from being sent to the candidate's spam folder, the candidate should add *exam@isaca.org* to his/her address book, whitelist or safe senders list.

In order to become CRISC-certified, candidates must pass the exam and must complete and submit an application for certification (and must receive confirmation from ISACA that the application is approved). The application is available on the ISACA web site at *www.isaca.org/criscapp*. Once the application is approved, the applicant will be sent confirmation of the approval. The candidate is not CRISC-certified, and cannot use the CRISC designation, until the candidate's application is approved. A processing fee of US $50 must accompany your CRISC application for certification.

The score report contains a subscore for each job practice domain. The subscores can be useful in identifying those areas in which the candidate may need further study before retaking the exam. Unsuccessful candidates should note that taking either a simple or weighted average of the subscores does not derive the total scaled score. Candidates receiving a failing score on the exam may request a rescoring of their answer sheet. This procedure ensures that no stray marks, multiple responses or other conditions interfered with computer scoring. Candidates should understand, however, that all scores are subjected to several quality control checks before they are reported; therefore, rescores most likely will not result in a score change. Requests for hand scoring must be made in writing to the certification department within 90 days following the release of the exam results. Requests for a hand score after the deadline date will not be processed. All requests must include a candidate's name, exam identification number and mailing address. A fee of US $75 must accompany this request.

GLOSSARY

A

Access control—The processes, rules and deployment mechanisms that control access to information systems, resources and physical access to premises

Access rights—The permission or privileges granted to users, programs or workstations to create, change, delete or view data and files within a system, as defined by rules established by data owners and the information security policy

Accountability—The ability to map a given activity or event back to the responsible party

Advanced persistent threat (APT)—An adversary that possesses sophisticated levels of expertise and significant resources which allow it to create opportunities to achieve its objectives using multiple attack vectors (NIST SP800-61)

> **Scope notes:** The APT:
> 1. Pursues its objectives repeatedly over an extended period of time
> 2. Adapts to defenders' efforts to resist it
> 3. Is determined to maintain the level of interaction needed to execute its objectives

Application controls—The policies, procedures and activities designed to provide reasonable assurance that objectives relevant to a given automated solution (application) are achieved

Architecture—Description of the fundamental underlying design of the components of the business system, or of one element of the business system (e.g., technology), the relationships among them, and the manner in which they support enterprise objectives

Asset—Something of either tangible or intangible value worth protecting, including people, information, infrastructure, finances and reputation

Asset value—The value of an asset is subject many factors including the value to both the business and to competitors. An asset may be valued according to what another person would pay for it, or by its measure of value to the company. Asset value is usually done using a quantitative (monetary) value.

Authentication—1. The act of verifying identity, i.e., user, system

> **Scope notes:** Risk: Can also refer to the verification of the correctness of a piece of data.

> 2. The act of verifying the identity of a user, the user's eligibility to access computerized information.

> **Scope notes:** Assurance: Authentication is designed to protect against fraudulent logon activity. It can also refer to the verification of the correctness of a piece of data.

Authenticity—Undisputed authorship

Availability—Ensuring timely and reliable access to and use of information

Awareness—Being acquainted with, mindful of, conscious of and well informed on a specific subject, which implies knowing and understanding a subject and acting accordingly

B

Balanced scorecard (BSC)—Developed by Robert S. Kaplan and David P. Norton as a coherent set of performance measures organized into four categories that includes traditional financial measures, but adds customer, internal business process, and learning and growth perspectives

Business case—Documentation of the rationale for making a business investment, used both to support a business decision on whether to proceed with the investment and as an operational tool to support management of the investment through its full economic life cycle

Business continuity—Preventing, mitigating and recovering from disruption

> **Scope notes:** The terms "business resumption planning," "disaster recovery planning" and "contingency planning" also may be used in this context; they focus on recovery aspects of continuity, and for that reason the 'resilience' aspect should also be taken into account.

Business continuity plan (BCP)—A plan used by an enterprise to respond to disruption of critical business processes. Depends on the contingency plan for restoration of critical systems

Business goal—The translation of the enterprise's mission from a statement of intention into performance targets and results

Business impact—The net effect, positive or negative, on the achievement of business objectives

Business impact analysis/assessment—Evaluating the criticality and sensitivity of information assets. An exercise that determines the impact of losing the support of any resource to an enterprise, establishes the escalation of that loss over time, identifies the minimum resources needed to recover, and prioritizes the recovery of processes and the supporting system.

> **Scope notes:** This process also includes addressing:
> - Income loss
> - Unexpected expense
> - Legal issues (regulatory compliance or contractual)
> - Interdependent processes
> - Loss of public reputation or public confidence

Business objective—A further development of the business goals into tactical targets and desired results and outcomes

Business process owner—The individual responsible for identifying process requirements, approving process design and managing process performance

> **Scope note:** Must be at an appropriately high level in the enterprise and have authority to commit resources to process-specific risk management activities

Business risk—A probable situation with uncertain frequency and magnitude of loss (or gain)

C

Capability—An aptitude, competency or resource that an enterprise may possess or require at an enterprise, business function or individual level that has the potential, or is required, to contribute to a business outcome and to create value

Capability Maturity Model (CMM)—1. Contains the essential elements of effective processes for one or more disciplines. It also describes an evolutionary improvement path from *ad hoc*, immature processes to disciplined, mature processes with improved quality and effectiveness

2. CMM for software, from the Software Engineering Institute (SEI), is a model used by many enterprises to identify best practices useful in helping them assess and increase the maturity of their software development processes

Scope notes: CMM ranks software development enterprises according to a hierarchy of five process maturity levels. Each level ranks the development environment according to its capability of producing quality software. A set of standards is associated with each of the five levels. The standards for level one describe the most immature or chaotic processes and the standards for level five describe the most mature or quality processes. A maturity model that indicates the degree of reliability or dependency the business can place on a process achieving the desired goals or objectives. A collection of instructions that an enterprise can follow to gain better control over its software development process

Change management—A holistic and proactive approach to managing the transition from a current to a desired organizational state, focusing specifically on the critical human or "soft" elements of change

Scope notes: Includes activities such as culture change (values, beliefs and attitudes), development of reward systems (measures and appropriate incentives), organizational design, stakeholder management, human resources (HR) policies and procedures, executive coaching, change leadership training, team building and communications planning and execution

Cloud computing—Convenient, on-demand network access to a shared pool of resources that can be rapidly provisioned and released with minimal management effort or service provider interaction

Compensating control—An internal control that reduces the risk of an existing or potential control weakness resulting in errors and omissions

Computer emergency response team (CERT)—A group of people integrated at the enterprise with clear lines of reporting and responsibilities for standby support in case of an information systems emergency

This group will act as an efficient corrective control, and should also act as a single point of contact for all incidents and issues related to information systems.

Confidentiality—Preserving authorized restrictions on access and disclosure, including means for protecting privacy and proprietary information

Configuration management—The control of changes to a set of configuration items over a system life cycle

Control—The means of managing risk, including policies, procedures, guidelines, practices or organizational structures, which can be of an administrative, technical, management, or legal nature

Control risk—The risk that a material error exists that would not be prevented or detected on a timely basis by the system of internal controls (See *Inherent risk*).

Control risk self-assessment—A method/process by which management and staff of all levels collectively identify and evaluate risk and controls with their business areas. This may be under the guidance of a facilitator such as an auditor or risk manager.

Copyright—Protection of writings, recordings or other ways of expressing an idea. The idea itself may be common, but they way it was expressed is unique, such as a song or book

Culture—A pattern of behaviors, beliefs, assumptions, attitudes and ways of doing things

Scope notes: COBIT 5 perspective

D

Data classification—The assignment of a level of sensitivity to data (or information) that results in the specification of controls for each level of classification. Levels of sensitivity of data are assigned according to predefined categories as data are created, amended, enhanced, stored or transmitted. The classification level is an indication of the value or importance of the data to the enterprise.

Data classification scheme—An enterprise scheme for classifying data by factors such as criticality, sensitivity and ownership

Data custodian—The individual(s) and department(s) responsible for the storage and safeguarding of computerized data

Data owner—The individual(s), normally a manager or director, who has responsibility for the integrity, accurate reporting and use of computerized data

Demilitarized zone (DMZ)—A screened (firewalled) network segment that acts as a buffer zone between a trusted and untrusted network

    Scope notes: A DMZ is typically used to house systems such as web servers that must be accessible from both internal networks and the Internet

Detective control— Exists to detect and report when errors, omissions and unauthorized uses or entries occur

Disaster recovery plan (DRP)—A set of human, physical, technical and procedural resources to recover, within a defined time and cost, an activity interrupted by an emergency or disaster

E

Encryption—The process of taking an unencrypted message (plaintext), applying a mathematical function to it (encryption algorithm with a key) and producing an encrypted message (ciphertext)

Encryption algorithm—A mathematically based function or calculation that encrypts/decrypts data

Enterprise resource planning (ERP) system—A packaged business software system that allows an enterprise to automate and integrate the majority of its business processes, share common data and practices across the entire enterprise, and produce and access information in a real-time environment

    Scope notes: Examples of ERP include SAP, Oracle Financials and J.D. Edwards.

Enterprise risk management (ERM)—The discipline by which an enterprise in any industry assesses, controls, exploits, finances and monitors risk from all sources for the purpose of increasing the enterprise's short- and long-term value to its stakeholders

Event—Something that happens at a specific place and/or time

Event type—For the purpose of IT risk management, one of three possible sorts of events: threat event, loss event and vulnerability event

    Scope notes: Being able to consistently and effectively differentiate the different types of events that contribute to risk is a critical element in developing good risk-related metrics and well-informed decisions. Unless these categorical differences are recognized and applied, any resulting metrics lose meaning and, as a result, decisions based on those metrics are far more likely to be flawed.

Evidence—1. Information that proves or disproves a stated issue

        2. Information an auditor gathers in the course of performing an IS audit; relevant if it pertains to the audit objectives and has a logical relationship to the findings and conclusions it is used to support

    Scope note: Audit perspective

F

Fallback procedures—A plan of action or set of procedures to be performed if a system implementation, upgrade or modification does not work as intended

    Scope note: May involve restoring the system to its state prior to the implementation or change. Fallback procedures are needed to ensure that normal business processes continue in the event of failure and should always be considered in system migration or implementation

Feasibility study—A phase of a system development life cycle (SDLC) methodology that researches the feasibility and adequacy of resources for the development or acquisition of a system solution to a user need

Framework—A generally accepted, business-process-oriented structure that establishes a common language and enables repeatable business processes

    Scope note: This term may be defined differently in different disciplines. This definition suits the purposes of this manual.

Frequency—A measure of the rate by which events occur over a certain period of time

G

Governance—Ensures that stakeholder needs, conditions and options are evaluated to determine balanced, agreed-on enterprise objectives to be achieved; setting direction through prioritization and decision making; and monitoring performance and compliance against agreed-on direction and objectives

    Scope note: Conditions can include the cost of capital, foreign exchange rates, etc. Options can include shifting manufacturing to other locations, subcontracting portions of the enterprise to third parties, selecting a product mix from many available choices, etc.

Governance enabler—Something (tangible or intangible) that assists in the realization of effective governance

    Scope notes: COBIT 5 perspective

Governance of enterprise IT—A governance view that ensures that information and related technology support and enable the enterprise strategy and the achievement of enterprise objectives; this also includes the functional governance of IT, i.e., ensuring that IT capabilities are provided efficiently and effectively.

    Scope notes: COBIT 5 perspective

I

Impact—Magnitude of loss resulting from a threat exploiting a vulnerability

Impact analysis—A study to prioritize the criticality of information resources for the enterprise based on costs (or consequences) of adverse events

In an impact analysis, threats to assets are identified and potential business losses determined for different time periods. This assessment is used to justify the extent of safeguards that are required and recovery time frames. This analysis is the basis for establishing the recovery strategy.

Impact assessment—A review of the possible consequences of a risk

    Scope notes: See also *Impact analysis*

Incident—Any event that is not part of the standard operation of a service and that causes, or may cause, an interruption to, or a reduction in, the quality of that service

Information security— Ensures that within the enterprise, information is protected against disclosure to unauthorized users (confidentiality), improper modification (integrity), and non-access when required (availability)

Information systems (IS)—The combination of strategic, managerial and operational activities involved in the gathering, processing, storing, distributing and use of information, and its related technologies

   Scope notes: Information systems are distinct from information technology (IT) in that an information system has an IT component that interacts with the process components

Information technology (IT)—The hardware, software, communication and other facilities used to input, store, process, transmit and output data in whatever form

Infrastructure as a Service (IaaS)—Offers the capability to provision processing, storage, networks and other fundamental computing resources, enabling the customer to deploy and run arbitrary software, which can include operating systems (OSs) and applications

Inherent risk—The risk level or exposure without taking into account the actions that management has taken or might take (e.g., implementing controls)

Integrity— The guarding against improper information modification or destruction, and includes ensuring information non-repudiation and authenticity

Intellectual property—Intangible assets that belong to an enterprise for its exclusive use

Internal controls—The policies, procedures, practices and organizational structures designed to provide reasonable assurance that the business objectives will be achieved and undesired events will be prevented or detected and corrected

IT architecture— Description of the fundamental underlying design of the IT components of the business, the relationships among them, and the manner in which they support the enterprise's objectives

IT infrastructure—The set of hardware, software and facilities that integrates an enterprise's IT assets

   Scope note: Specifically, the equipment (including servers, routers, switches, and cabling), software, services and products used in storing, processing, transmitting and displaying all forms of information for the organization's users

IT-related incident—An IT-related event that causes an operational, developmental and/or strategic business impact

IT risk—The business risk associated with the use, ownership, operation, involvement, influence and adoption of IT within an enterprise

IT risk issue—1. An instance of an IT risk

   2. A combination of control, value and threat conditions that impose a noteworthy level of IT risk

IT risk profile—A description of the overall (identified) IT risk to which the enterprise is exposed

IT risk register—A repository of the key attributes of potential and known IT risk issues. Attributes may include name, description, owner, expected/actual frequency, potential/actual magnitude, potential/actual business impact and disposition

IT risk scenario—The description of an IT-related event that can lead to a business impact

IT strategic plan—A long-term plan (i.e., three- to five-year horizon) in which business and IT management cooperatively describe how IT resources will contribute to the enterprise's strategic objectives (goals)

IT tactical plan—A medium-term plan (i.e., six- to 18-month horizon) that translates the IT strategic plan direction into required initiatives, resource requirements and ways in which resources and benefits will be monitored and managed

K

Key performance indicator (KPI)—A measure that determines how well the process is performing in enabling the goal to be reached

> **Scope Note:** A lead indicator of whether a goal will likely be reached, and a good indicator of capabilities, practices and skills. It measures an activity goal, which is an action that the process owner must take to achieve effective process performance

Key risk indicator (KRI)—A subset of risk indicators that are highly relevant and possess a high probability of predicting or indicating important risk

> **Scope note:** See *Risk indicator*

L

Likelihood—The probability of something happening

Loss event— Any event where a threat event results in loss

> **Scope note:** From Jones, J.; "FAIR Taxonomy," Risk Management Insight, USA, 2008

M

Magnitude— A measure of the potential severity of loss or the potential gain from realized events/scenarios

Management—Plans, builds, runs and monitors activities in alignment with the direction set by the governance body to achieve the enterprise objectives

N

Nondisclosure agreement (NDA)— A legal contract between at least two parties that outlines confidential materials that the parties wish to share with one another for certain purposes, but wish to restrict from generalized use; a contract through which the parties agree not to disclose information covered by the agreement

> **Scope notes:** Also called a confidential disclosure agreement (CDA), confidentiality agreement or secrecy agreement. An NDA creates a confidential relationship between the parties to protect any type of trade secret. As such, an NDA can protect non-public business information. In the case of certain governmental entities, the confidentiality of information other than trade secrets may be subject to applicable statutory requirements, and in some cases may be required to be revealed to an outside party requesting the information. Generally, the governmental entity will include a provision in the contract to allow the seller to review a request for information that the seller identifies as confidential and the seller may appeal such a decision requiring disclosure. NDAs are commonly signed when two companies or individuals are considering doing business together and need to understand the processes used in one another's businesses solely for the purpose of evaluating the potential business relationship. NDAs can be "mutual," meaning that both parties are restricted in their use of the materials provided, or they can only restrict a single party. It is also possible for an employee to sign an NDA or NDA-like agreement with a company at the time of hiring; in fact, some employment agreements will include a clause restricting "confidential information" in general.

Nonrepudiation—The assurance that a party cannot later deny originating data; provision of proof of the integrity and origin of the data and that can be verified by a third party

> **Scope notes:** A digital signature can provide nonrepudiation

O

Objectivity—The ability to exercise judgment, express opinions and present recommendations with impartiality

Operational level agreement (OLA)—An internal agreement covering the delivery of services that support the IT organization in its delivery of services

Owner— Individual or group that holds or possesses the rights of and the responsibilities for an enterprise, entity or asset

> **Scope notes:** Examples: process owner, system owner

P

Patent—Protection of research and ideas that led to the development of a new, unique and useful product to prevent the unauthorized duplication of the patented item

Penetration testing—A live test of the effectiveness of security defenses through mimicking the actions of real-life attackers

Performance indicators—A set of metrics designed to measure the extent to which performance objectives are being achieved on an ongoing basis

> **Scope note:** Performance indicators can include service level agreements (SLA), critical success factors (CSF), customer satisfaction ratings, internal or external benchmarks, industry best practices and international standards

Platform as a Service (PaaS)—Offers the capability to deploy onto the cloud infrastructure customer-created or -acquired applications that are created using programming languages and tools supported by the provider

Policy—1. Generally, a document that records a high-level principle or course of action that has been decided on

The intended purpose is to influence and guide both present and future decision making to be in line with the philosophy, objectives and strategic plans established by the enterprise's management teams.

> **Scope notes:** In addition to policy content, policies need to describe the consequences of failing to comply with the policy, the means for handling exceptions, and the manner in which compliance with the policy will be checked and measured.

2. Overall intention and direction as formally expressed by management

> **Scope notes:** COBIT 5 perspective

Portfolio—A grouping of "objects of interest" (investment programs, IT services, IT projects, other IT assets or resources) managed and monitored to optimize business value. (The investment portfolio is of primary interest to Val IT. The IT service, project, asset and other resource portfolios are of primary interest to COBIT.)

Preventive control—An internal control that is used to avoid undesirable events, errors and other occurrences that an enterprise has determined could have a negative material effect on a process or end product

Privilege—The level of trust with which a system object is imbued

Problem—In IT, the unknown underlying cause of one or more incidents

Problem escalation procedure—The process of escalating a problem up from junior to senior support staff, and ultimately to higher levels of management

> **Scope notes:** Problem escalation procedure is often used in help desk management, when an unresolved problem is escalated up the chain of command, until it is solved.

Program—A structured grouping of interdependent projects that is both necessary and sufficient to achieve a desired business outcome and create value. These projects could include, but are not limited to, changes in the nature of the business, business processes and the work performed by people as well as the competencies required to carry out the work, the enabling technology and the organizational structure.

Project—A structured set of activities concerned with delivering a defined capability (that is necessary, but not sufficient, to achieve a required business outcome) to the enterprise, based on an agreed-on schedule and budget

Project portfolio—The set of projects owned by a company

   Scope note: It usually includes the main guidelines relative to each project, including objectives, costs, time lines and other information specific to the project

Q

Qualitative risk analysis— Defines risk using a scale or comparative values (i.e., defining risk factors in terms of high/medium/low or on a numeric scale from 1 to 10). It is based on judgment, intuition and experience rather than on financial values

Quantitative risk analysis— The use of numerical and statistical techniques to calculate likelihood and impact of risk. It uses financial data, percentages and ratios to provide an approximate measure of the magnitude of impact in financial terms

R

RACI chart—Illustrates who is Responsible, Accountable, Consulted and Informed within an organizational framework

Recovery point objective (RPO)—Determined based on the acceptable data loss in case of a disruption of operations. It indicates the earliest point in time to which it is acceptable to recover the data. The RPO effectively quantifies the permissible amount of data loss in case of interruption.

Recovery strategy—An approach by an enterprise that will ensure its recovery and continuity in the face of a disaster or other major outage

   Scope notes: Plans and methodologies are determined by the enterprise's strategy. There may be more than one methodology or solution for an enterprise's strategy. Examples of methodologies and solutions include contracting for hot site or cold site, building an internal hot site or cold site, identifying an alternate work area, a consortium or reciprocal agreement, contracting for mobile recovery or crate and ship, and many others.

Recovery testing—A test to check the system's ability to recover after a software or hardware failure

Recovery time objective (RTO)—The amount of time allowed for the recovery of a business function or resource after a disaster occurs

Residual risk—The remaining risk after management has implemented risk response

Resilience—The ability of a system or network to resist failure or to recover quickly from any disruption, usually with minimal recognizable effect

Return on investment (ROI)—A measure of operating performance and efficiency, computed in its simplest form by dividing net income by the total investment over the period being considered

Risk—The combination of the probability of an event and its consequence. (ISO/IEC73)

Risk acceptance—If the risk is within the enterprise's risk tolerance or if the cost of otherwise mitigating the risk is higher than the potential loss, the enterprise can assume the risk and absorb any losses

Risk aggregation—The process of integrating risk assessments at a corporate level to obtain a complete view on the overall risk for the enterprise

Risk analysis—1. A process by which frequency and magnitude of IT risk scenarios are estimated.

2. The initial steps of risk management: analyzing the value of assets to the business, identifying threats to those assets and evaluating how vulnerable each asset is to those threats

Scope notes: It often involves an evaluation of the probable frequency of a particular event, as well as the probable impact of that event

Risk appetite—The amount of risk, on a broad level, that an entity is willing to accept in pursuit of its mission

Risk assessment—A process used to identify and evaluate risk and its potential effects

Scope note: Includes assessing the critical functions necessary for an enterprise to continue business operations, defining the controls in place to reduce organization exposure and evaluating the cost for such controls. Risk analysis often involves an evaluation of the probabilities of a particular event.

Risk avoidance—The process for systematically avoiding risk, constituting one approach to managing risk

Risk culture—The set of shared values and beliefs that governs attitudes toward risk-taking, care and integrity, and determines how openly risk and losses are reported and discussed

Risk evaluation—The process of comparing the estimated risk against given risk criteria to determine the significance of the risk (ISO/IEC Guide 73:2002)

Risk factor—A condition that can influence the frequency and/or magnitude and, ultimately, the business impact of IT-related events/scenarios

Risk identification—The process of determining and documenting the risk that an enterprise faces. The identification of risk is based on the recognition of threats, vulnerabilities, assets and controls in the enterprise's operational environment

Risk impact—The calculation of the amount of loss or damage that an organization may incur due to a risk event

Risk indicator—A metric capable of showing that the enterprise is subject to, or has a high probability of being subject to, a risk that exceeds the defined risk appetite

Risk management—1. The coordinated activities to direct and control an enterprise with regard to risk

Scope notes: In the International Standard, the term "control" is used as a synonym for "measure" (ISO/IEC Guide 73:2002)

2. One of the governance objectives. Entails recognizing risk; assessing the impact and likelihood of that risk; and developing strategies, such as avoiding the risk, reducing the negative effect of the risk and/or transferring the risk, to manage it within the context of the enterprise's risk appetite

Scope notes: COBIT 5 perspective

Risk map—A (graphic) tool for ranking and displaying risk by defined ranges for frequency and magnitude

Risk mitigation—The management of risk through the use of countermeasures and controls

Risk portfolio view—1. A method to identify interdependencies and interconnections among risk, as well as the effect of risk responses on multiple types of risk

        2. A method to estimate the aggregate impact of multiple types of risk (e.g., cascading and coincidental threat types/scenarios, risk concentration/correlation across silos) and the potential effect of risk response across multiple types of risk

Risk scenario—A description of an event that can lead to a business impact

Risk tolerance—The acceptable level of variation that management is willing to allow for any particular risk as the enterprise pursues its objectives

Risk transfer—The process of assigning risk to another enterprise, usually through the purchase of an insurance policy or by outsourcing the service

Root cause analysis—A process of diagnosis to establish the origins of events, which can be used for learning from consequences, typically from errors and problems

S

Scope creep—Also called requirement creep, this refers to uncontrolled changes in a project's scope

    Scope notes: Scope creep can occur when the scope of a project is not properly defined, documented and controlled. Typically, the scope increase consists of either new products or new features of already approved products. Hence, the project team drifts away from its original purpose. Because of one's tendency to focus on only one dimension of a project, scope creep can also result in a project team overrunning its original budget and schedule. For example, scope creep can be a result of poor change control, lack of proper identification of what products and features are required to bring about the achievement of project objectives in the first place, or a weak project manager or executive sponsor

Segregation/separation of duties (SoD)—A basic internal control that prevents or detects errors and irregularities by assigning to separate individuals the responsibility for initiating and recording transactions and for the custody of assets

    Scope notes: Segregation/separation of duties is commonly used in large IT organizations so that no single person is in a position to introduce fraudulent or malicious code without detection

Service level agreement (SLA)—An agreement, preferably documented, between a service provider and the customer(s)/user(s) that defines minimum performance targets for a service and how they will be measured

Slack time (float)—Time in the project schedule, the use of which does not affect the project's critical path; the minimum time to complete the project based on the estimated time for each project segment and their relationships

    Scope note: Slack time is commonly referred to as "float" and generally is not "owned" by either party to the transaction

Software as a Service (SaaS)—Offers the capability to use the provider's applications running on cloud infrastructure. The applications are accessible from various client devices through a thin client interface such as a web browser (e.g., web-based email)

Standard—A mandatory requirement, code of practice or specification approved by a recognized external standards organization, such as the International Organization for Standardization (ISO)

Statement of work (SOW)—A formal document that captures and defines the work activities, deliverables, and time line a vendor must execute in performance of specified work for a client

The SOW usually includes detailed requirements and pricing, with standard regulatory and governance terms and conditions

Strategic planning—The process of deciding on the enterprise's objectives, on changes in these objectives, and the policies to govern their acquisition and use

System development life cycle (SDLC)—The phases deployed in the development or acquisition of a software system

> **Scope notes:** SDLC is an approach used to plan, design, develop, test and implement an application system or a major modification to an application system. Typical phases of SDLC include the feasibility study, requirements study, requirements definition, detailed design, programming, testing, installation and post-implementation review, but not the service delivery or benefits realization activities.

T

Threat—Anything (e.g., object, substance, human) that is capable of acting against an asset in a manner that can result in harm

> **Scope note:** A potential cause of an unwanted incident (ISO/IEC 13335)

Threat agent—Methods and things used to exploit a vulnerability

> **Scope notes:** Examples include determination, capability, motive and resources

Threat analysis—An evaluation of the type, scope and nature of events or actions that can result in adverse consequences; identification of the threats that exist against enterprise assets

> **Scope notes:** The threat analysis usually defines the level of threat and the likelihood of it materializing

Threat event—Any event where a threat element/actor acts against an asset in a manner that has the potential to directly result in harm

Threat vector—The path or route used by the adversary to gain access to the target

Trademark—A sound, color, logo, saying or other distinctive symbol that is closely associated with a certain product or company

V

Vulnerability—A weakness in the design, implementation, operation or internal control of a process that could expose the system to adverse threats from threat events

Vulnerability analysis—A process of identifying and classifying vulnerabilities

Vulnerability event—Any event where a material increase in vulnerability results. Note that this increase in vulnerability can result from changes in control conditions or from changes in threat capability/force

> **Scope notes:** From Jones, J.; "FAIR Taxonomy," Risk Management Insight, USA, 2008

Vulnerability scanning—An automated process to proactively identify security weaknesses in a network or individual system

Index

A slash (/) indicates that the terms are synonymous within this manual.

"See also" indicates that the terms are related or relevant to one another.

W

WAN, See Wide area network
Wide area network (WAN), 94-96
Wireless Access Points, 94

EVALUATION

ISACA continuously monitors the swift and profound professional, technological and environmental advances affecting risk and IS control professionals. Recognizing these rapid advances, the *CRISC™ Review Manual* is updated annually.

To assist ISACA in keeping abreast of these advances, please take a moment to evaluate the *CRISC™ Review Manual 2015*. Such feedback is valuable to fully serve the profession and future CRISC exam registrants.

To complete the evaluation on the web site, please go to *www.isaca.org/studyaidsevaluation.*

Thank you for your support and assistance.

Prepare for the
2015 CRISC Exams

2015 CRISC Review Resources for Exam Preparation and Professional Development

Successful Certified in Risk and Information Systems Control™ (CRISC™) exam candidates have an organized plan of study. To assist individuals with the development of a successful study plan, ISACA® offers several study aids and review courses to exam candidates. These include:

Study Aids

- *CRISC™ Review Manual 2015*

- *CRISC™ Review Questions, Answers & Explanations Manual 2015*

- *CRISC™ Review Questions, Answers & Explanations Manual 2015 Supplement*

- CRISC™ Review Questions, Answers & Explanations Database -12 Month Subscription

To order, visit *www.isaca.org/criscbooks*.

Review Courses

- Chapter-sponsored review courses *(www.isaca.org/criscreview)*

Glossary for CRISC